Nature Inside

Nature Inside

Plants and Flowers
in the Modern Interior

Penny Sparke

Yale University Press
New Haven and London

First published by Yale University Press 2020
302 Temple Street, P.O. Box 209040,
New Haven CT 06520-9040
47 Bedford Square, London WC1B 3DP
yalebooks.com | yalebooks.co.uk

Copyright © 2020 Penny Sparke

All rights reserved. This book may not be reproduced or transmitted in any form or by any means, electronic or mechanical, including photocopy, recording or any other information storage and retrieval system (beyond that copying permitted by Sections 107 and 108 of the US Copyright Law and except by reviewers for the public press), without prior permission in writing from the publisher.

ISBN 978-0-300-24402-1 HB
Library of Congress Control Number: 2020931947

10 9 8 7 6 5 4 3 2 1
2024 2023 2022 2021 2020

Designed by Emma Kalkhoven

Printed in China

Front cover: Eames House (Case Study House 8), Los Angeles, designed by Charles and Ray Eames, 1949. Photograph by Julius Shulman, 1950. © J. Paul Getty Trust. Getty Research Institute, Los Angeles (2004.R.10)

Contents

7 *Acknowledgements*

9 INTRODUCTION

15 CHAPTER 1
Taming the Jungle

33 CHAPTER 2
The Jungle in the Parlour

51 CHAPTER 3
The Parlour Outside

69 CHAPTER 4
Natural Modernism

87 CHAPTER 5
The Garden Inside:
The Villa Tugendhat, Brno

105 CHAPTER 6
Living in the Garden:
Californian Modernism

125 CHAPTER 7
Natural Late Modernism

143 CHAPTER 8
The Living Room in the City:
The Hyatt Regency Hotel, Atlanta

163 CHAPTER 9
The Benefits of Nature Inside

181 CHAPTER 10
Greening the Interior

200 *Notes*
204 *Bibliography*
216 *Index*
224 *Picture Credits*

Acknowledgements

This publication has been made possible by grants from the Graham Foundation for Advanced Studies in the Fine Arts; the Scouloudi Foundation in association with the Institute of Historical Research; and the Society of Architectural Historians of Great Britain. I would also like to thank Oliver Heath Design for providing me with images of their work.

Other thanks go to a vast number of people, too many to all be named here, who provided much-welcomed help and support during the extended research and writing process of *Nature Inside* in a variety of ways. They include: Jeremy Aynsley; Chris Barrett; Tim Benton; Pat Brown; Alison Clarke; Hilary Dalke; Leo Duff; David Falkner; Fiona Fisher; Hilary French; Alice Friedman; Christine Guth; Kristina Hansen Hadberg; Ian Higgins; Lisa Hirst; Lisa Hockemeyer; Ersi Ioannadou; Pat Kirkham; Sue Knights; Patricia Lara-Betancourt; Anca Lasc; Sarah Lichtman; Paula Lupkin; Sorcha O'Brien; Clare O'Mahony; Anders Munch; Jane Pavitt; John Potvin; Rebecca Preston; Charles Rice; Cat Rossi; Jennifer Salahub; Joel Sanders; Catherine Sidwell; Alex Suarez; Jim Sullivan; Vanessa van den Berghe; Andy Wallace; Suzette Worden; and James Woudhuysen.

I would like to dedicate the book to the memories of Peter Gallimore, who introduced me to Samuel J. Waring's house, Palmyra, in Liverpool, and Sandra Alfoldy, at whose conference – NeoCraft, held in Halifax, Nova Scotia, in 2007 – I first thought of the subject of *Nature Inside*.

At Yale University Press I am indebted to Gillian Malpass, who was first enthused by the idea of *Nature Inside*, Mark Eastment, Julie Hrischeva, Sophie Neve, Clare Davis, Marianne Fisher and Emma Kalkhoven.

Finally, special thanks go to my three daughters, Molly, Nancy and Celia, who accompanied me on several research trips. They are among the best research assistants and bag carriers that I know, and they supported me in many different ways throughout the whole project.

Introduction

Study nature, love nature, stay close to nature. It will never fail you.[1]
Frank Lloyd Wright

We seem, today, to be infatuated with indoor plants and flowers. Many of our everyday public indoor spaces are full of them, while glossy magazines depict miniature jungles in stylish domestic interiors. In 2011, six fully grown English field maples and two hornbeams formed part of the lavish setting that framed Prince William and Catherine Middleton's wedding in Westminster Abbey, while the trio of glass orbs that form Amazon's headquarters in Seattle contain over 40,000 plants (opened 2018; fig. 1). In addition, countless health and leisure centres, hotel lobbies, exhibition halls, corporate offices, railway stations, airports and shopping malls across the globe house elaborate indoor planting schemes.

Putting plants and flowers in containers and bringing them inside are not new activities, however. Egyptian, early Chinese and Middle Eastern civilisations embraced such practices, as did the ancient Greeks. The Romans grew plants in terracotta pots and filled the inner atria of their houses with them, though those spaces were open to the sky.[2] Yet despite these continuities, by the late eighteenth century nature inside had taken on multiple new meanings in the Western world. This book sets out to tell a story about the inclusion of plants and flowers in interior spaces from that moment onwards, with an emphasis on developments in Britain, continental Europe and the United States. Through its emphasis on nature inside it aims to bring together into a single narrative a study of nineteenth-century domestic interiors and public winter gardens, with analyses of the interiors of canonical (and less canonical) examples of inter-war modernist, post-1945 late modernist and contemporary architecture.[3]

Fig. 1 Inside one of the giant biospheres at Amazon's headquarters in Seattle, 2018. Photograph by Mike Kane

Above all, this new narrative is characterised by continuity rather than, as is usually presented, by the disruptive shift from nineteenth-century historicism and eclecticism to modernist minimalism and beyond. 'Modernity provokes on all levels an aesthetic of rupture', the French cultural theorist Jean Baudrillard has written.[4] In this book, by contrast, numerous twentieth- and twenty-first-century interiors are unpacked in the context of nineteenth-century ideas and practices related to bringing nature inside. While the functions and meanings of indoor nature have inevitably changed over time, its roles as an *aide-memoire* of the pre-industrial past, as a form of therapy for human beings, and as an active agent in the creation of a non-toxic environment have arguably remained intact through the whole period. Providing a counterpoint to rapid urbanisation, to the hard forms and materials of modern architecture, and to the cultural dominance of advanced technologies, in a variety of ways nature inside has consistently offered a much-needed anchor to otherwise unchained 'progress'.

Historians of interiors rarely address the aspidistra in the corner, or the vase of flowers on the table. Plants and flowers grow and die, and are often thought to have been put in place at the last minute for photographers. Ignored as vital components of spaces, they are relegated to the inferior status of representations of nature, mere onlookers in the interiors they occupy. On one level, therefore, this book offers a new account of the history of modern architecture and interior design that focuses on the presence of plants and flowers in otherwise exclusively cultural settings. Given that this is a design-historical study, the structure of the book follows a familiar chronological narrative that moves through Victorian historicism to European modernism and on to American and global late modernism. This time, however, that well-trodden path is viewed through a completely new lens.

The book begins by explaining that nature inside first emerged in the modern Western world in the context of the overseas trade and the acquisition of plants and flowers from European colonies that took place from the seventeenth to the nineteenth centuries. As increasing numbers of people became urban dwellers, however, the established relationship between nature and human beings reached a crisis. The deliberate rejection of nature that came with increasing industrialisation brought with it a reaction in the form of romanticism, which manifested at a high cultural level in poetry and music, and was represented in a more popular form by, among other phenomena, the desire to bring nature, in the shape of plants and flowers, inside. Quite quickly, plants' and flowers' newly acquired messages about taste, fashion and respectable middle-class domesticity were also injected into the public sphere, where they were often used instrumentally to conceal rampant commercialism.

In the inter-war years of the twentieth century a sharp shift occurred away from nature inside's integration into popular culture and links to nineteenth-century modernity. Instead, it took on a new role within the radical ideas about architecture and interiors which defined the high-cultural movement that emerged in the West, known today as modernism. In spite of that movement's declared rejection of domesticity, plants and flowers continued to play a significant part in the interior schemes of many modernist buildings. However, neither the architect-designers of the time nor the later historians of the movement acknowledged their important role.

In the years after 1945 the movement of architectural and design modernism across the Atlantic facilitated two important developments. First, due to the fine weather on North America's west coast, the Californian modernist architects who moved there from Europe realised fully what had been only a dream up to that point, namely the erosion of the boundary between residential architecture's inside and outside spaces; between, that is, nature and culture. Second, for reasons not unlike those which had drawn nature out of the private sphere in the late nineteenth century, as a result of the United States' post-war economic boom, many large public and semi-public late-modern interiors also began to embrace nature inside.

Finally, in the period from 1970 to the present, a desire to subject indoor plants and flowers to scientific scrutiny came to the fore. Evidence from this research was used to reassure the managers of commercial spaces that investing in indoor planting could contribute to their profits. The plantscaping profession emerged on the back of this. Plants and flowers have now penetrated indoor environments globally on an unprecedented scale, and their meanings have become inextricably entangled with the popular environmental movement.

Over the years nature inside has become, in turn, a commodity, a marker of colonial power, a sign of aristocratic wealth and status, a component of

Victorian middle-class feminine domesticity, an aesthetic strategy within European modernist architecture, a humanising factor within late modernism in the private and public spheres, a tool within late economic capitalism, a subject of the modern scientific paradigm and a marker of the environmental crisis.

This book uses many examples to illustrate its vast subject. Together they tell a story of the main trends and ideas that drove the movement of nature inside. Several of the individual plants used in the interior settings under review – including the palm, the fern, the cactus, the Swiss cheese plant and the rubber plant – have been singled out as representatives of the moments in which they were both fashionable and meaningful. The focus throughout the book is on real plants located in actual interiors rather than on the multiple ways in which nature has been represented in paintings, on wallpaper and on furnishing fabrics, or imagined in idealised settings. The aim has been to add the idea of the 'natural' to the categories of the visual, the material and the spatial.

In addition to the case made for historical continuity, which sits at the heart of this study, four important themes weave in and out of its narrative. First, the commodification of plants was a by-product of the colonial aims of classification, control and ownership. Also, without the trading that took place during the colonial era, most of the plants that were subsequently used in inside spaces would never have left their native lands. The legacy of colonialism is therefore felt throughout this story.

Second, an association between nature and the female gender occurs as a *leitmotif* throughout this study, especially, unsurprisingly, in the context of the domestic sphere. The facts that nineteenth-century greenhouses were usually attached to ladies' rooms; that the responsibility for nineteenth-century window gardening lay with female amateur interior decorators; that several female architects and designers, among them Lilly Reich, Aino Aalto and Ray Eames, were probably responsible for introducing nature into the interiors of buildings designed by their modernist husbands and partners; and that the ecofeminist movement, which emerged in the 1980s, has played such a key role within recent environmental politics, all support that association.[5]

The practice of adding plants and flowers to interiors began, in an upper-class setting, as an extension of (male) gardening. It subsequently moved into the hands of middle-class housewives where, in the context of the domestic sphere, it has largely remained. As the activity became widespread in the public arena it became the responsibility of a range of (predominantly male) professionals: architects, interior decorators and designers, landscape architects and, latterly, the more specialist interiorscapers and plantscapers. The book also sets out to recount that gendered story.

Thirdly, while plants and flowers were embraced in a range of private, semi-private and public indoor spaces throughout the period under review, the intentions behind their inclusion in those different arenas varied considerably.[6] In the private home, individuals benefited directly, both physically and psychologically, from the presence of indoor nature and its ability to absorb toxins from the air and oxygenate it. Once nature inside entered the commercial arena, however, it was frequently used to further the ends of economic capitalism. Human beings, it was believed, worked and shopped harder when they were surrounded by relaxing plants.

Finally, the idea that plants and flowers are active rather than passive inhabitants of inside spaces also pervades the pages of this book. In the final decades of the twentieth century the French philosopher Bruno Latour, together with members of the environmentalist movement, suggested that human beings are not alone in being able to exercise agency.[7] In 2004, claiming that 'plants can curry favour with their surrounding humans by showing what attractive displays they might create if given enough care', Russell Hitchings even suggested that, though human beings believe they control nature, the truth may be the reverse: that nature is controlling them.[8]

Although this book deals specifically with the story of indoor plants and flowers, it is also important to remember that they function as representatives of nature in its entirety. They are, therefore, implicated in the complex debates that have taken place over many years about the relationship between nature and culture. Since the eighteenth century these two concepts have been understood as having a dualistic relationship, one which many ecologically oriented writers see as lying at the heart of today's environmental issues.[9] Nature became a (near) component of culture through its commodification from the eighteenth century onwards.[10] Its presence in inside spaces – that is, in the human-constructed visual, material and spatial phenomena we call 'interiors' – brought it even closer to culture. As a result, plants and flowers became quasi-objects, contained and displayed in inside spaces. That double layering of cultural formation makes this study's task of deciphering the role of indoor plants and flowers a complex one.

Indoor plants and flowers are also emblematic of humanity's anxieties about the growing imbalance in its relationship with nature. That disharmony is a cumulative consequence of the control people have exerted over nature for centuries. Indeed, the strong concern that is felt about environmental issues in the early twenty-first century, and its eco-feminist implications, lay behind my decision to write about indoor plants and flowers in the first place.

While, on one level, this book adds a new dimension to the familiar histories of interior decoration and design from the late eighteenth century to the present, on another, it provides fresh insights into human beings' ever-changing relationship with the natural world over that period. It also acknowledges that all those responsible for bringing nature inside over the period under review largely understood, whether consciously or subconsciously, that indoor plants and flowers exert a level of agency. Finally, it lays the groundwork for the idea that, if human beings could recognise the importance of nature in an indoor setting, they might begin to develop a new respect for the entire natural environment.

CHAPTER 1

Taming the Jungle

Plants seldom figure in the grand narratives of war, peace, or even everyday life in proportion to their importance to humans. Yet they are significant natural and cultural artifacts, often at the center of high intrigue.[1]
Londa Schiebinger

The years between 1600 and 1914 – the era of modernity in the Western world – saw wild nature subordinated to the will, authority and desires of human beings. An important part of that subordination was represented by nature's physical containment in a range of inside spaces located in both the public and the private spheres. Several drivers propelled that taming of the jungle. Through the sixteenth, seventeenth and eighteenth centuries the curiosity, acquisitiveness and delight in exotic luxuries that underpinned Europe's desire to control overseas territories led to the importation of non-native natural species from the colonies. Taming nature both mitigated the fear of the 'other' and played an important part in the financial exploitation that was part and parcel of colonialism. While the practical applications of plants expanded over the period in question – herbs, drugs, spices, dyes, and a range of food, drinks and other products (such as sugar, tea, coffee and textiles) were traded widely – plants and flowers themselves also became desirable luxury commodities. They were nurtured by horticulturists in commercial nurseries and by botanists in botanical gardens, a process which assisted in their transformation from natural phenomena into (near) cultural artefacts.

Plant-hunting, which emerged in the seventeenth century, represented one branch of the commercial activity that defined European colonialism. Hunters sought to augment the supply of plants for the manufacture of the plant-related products listed above, to acquire specimens for scientific

Fig. 2 Portrait of Sir Hans Sloane, Irish physician and plant collector. Engraving by an unknown artist after an original painting by Stephen Slaughter, 1736, in the British Museum. 12.2 × 10 in.

research, to meet the requirements of the growing numbers of plant collectors, to provide plants for the many commercial nurseries that were setting up in business, and to increase the supply of exotic plants for wealthy landowners who were growing them on their estates.

Intrepid hunter-explorers were sent abroad by private collectors, kings, scientific academies and governments. Many of the early English hunters were physicians who worked on ships in that capacity. Notable among them was Sir Hans Sloane (fig. 2), the physician to Queen Anne and to kings George I and George II. Among his numerous botanical voyages, Sloane travelled to Jamaica in 1687, where he hunted for new drugs.[2] Physicians were soon joined by horticulturists and botanists. Sir Joseph Banks – a well-known naturalist and botanist who was president, from 1778, of the Royal Society, and an informal advisor to George III at the Royal Botanic Gardens, Kew, from 1797 – believed that horticulture and botany would benefit the nation financially.[3] By the middle of the eighteenth century trees were arriving in Britain from America, and by the end of it numerous plants from Africa, Australia and the Far East were also making an appearance.[4] By the early decades of the nineteenth century, growing exotic flowers and shrubs had become a possibility for many of Europe's gardeners.

The British East India Company, which was founded in 1599, and the Dutch East India Company, established three years later, played key roles in the transportation of plants and flowers from the colonies to Europe. The latter operated for two centuries from the Cape of Good Hope to Japan.

The Dutch West India Company, established in 1621, undertook its activities in West Africa, the Caribbean and North and South America. Their ships moved around the globe on new trade routes and supplied Europe with tea and silk from China, sugar and coffee from the West Indies, spices from the East Indies, and corn and tobacco from the American colonies.[5] By the early eighteenth century both seeds and living plants – including ericas, geraniums, pelargoniums, succulents and proteaceous plants from the Cape of South Africa – were also being brought to Europe on the trading ships.[6]

Before they were physically brought into the interior, the exotic plants and flowers that were transported to Europe were subjected to interventions by both scientists and artists. From the seventeenth century onwards the number of scientific publications increased in England and scientific work was facilitated by learned societies, among them the Royal Society, which was founded in 1660. The interest of horticulturists, botanists and taxonomists in plants as objects of scientific investigation represented a desire to assert the Enlightenment belief in the pre-eminence of reason over emotion and intuition.

One way in which the desire to subordinate nature to reason manifested was in the drive to formulate a system of classification, a taxonomy, that would provide shared names for previously unknown plants. As Londa Schiebinger has explained, 'Europe's naturalists not only collected the stuff of nature but lay their own peculiar grid of reason over nature so that nomenclatures and taxonomies often served as "tools of empire".'[7] Antonio Lafuente and Nuria Valverde reiterated her point when they wrote that 'Empire requires that scientists and their patrons share the belief that the stuff of nature can be captured in words, figures, lines, shading, gradients, or flows', reinforcing the idea that nature was controlled by the application of both reason and language.[8] The philosopher Val Plumwood has pointed out that naming something, as Christopher Columbus named the lands he discovered, is a way of claiming ownership and therefore of asserting control.[9]

The Swedish botanist Carl Linnaeus established a system whereby names of plants were authenticated by mounting dried plant specimens onto cards that were then stored in buildings called herbaria, many of which were linked to botanical gardens. In the years under review, those gardens gradually moved away from medicine and reoriented themselves towards scientific research. The French philosopher Michel Foucault has claimed that Linnaeus's work was crucial to the process of disciplining the hitherto 'unwieldy stuff of nature', while, according to Schiebinger, it constituted 'the starting point for modern botany in the twentieth century'.[10]

Aesthetic control over tamed plants and flowers was exerted by the British rural elite, who integrated them into the constructed landscapes on their estates (fig. 3). Taming nature by transforming it into aesthetically conceived gardens went back many centuries. By the eighteenth century the international focus was on Britain, where the so-called 'natural' garden

Fig. 3 The Temple of Apollo in the landscaped garden at Stourhead, Wiltshire. The garden was begun in 1741 by Henry Hoare II. Photograph by the author, 2013

had come to prominence, succeeding the earlier, more formal, European baroque garden.[11] The notion of the 'picturesque' also emerged at that time, an aesthetic ideal that was based on the concepts of beauty and the sublime as they were perceived by viewers moving through spatial settings. This celebration of the picturesque led to the ideas that the natural world was pleasing to human beings and that it exerted a form of agency which compelled humans to engage with it. In subsequent years, the picturesque aesthetic was also to underpin the ways in which tamed plants and flowers were arranged in indoor settings.

Many English gardens were designed and installed by a small number of hugely successful landscape gardeners, William Kent, Lancelot 'Capability' Brown and Humphry Repton among them. As well as creating gardens for Chiswick House in London, Kent also worked at Stowe House in Buckinghamshire and Rousham House in Oxfordshire, among other places. Brown's work included gardens at Petworth House in Sussex and at Tottenham Park in Wiltshire, while Repton was known for several garden designs including his work at Wembley Park in London, Blaise Castle near Bristol and Woburn Abbey, Bedfordshire. In the words of Andrea Wulf, by the eighteenth century landscape gardens had 'become the absolute Necessities of Life, without which a gentleman of the smallest fortune thinks he makes no Figure in the country'.[12] When the 3rd Earl of Burlington created his garden at Chiswick House, he proclaimed that 'botany and horticulture were unimportant compared with the architectural embellishment of the garden'.[13] Sometimes artistically oriented landscape gardening could be combined with scientific enquiry, as wealthy landowners, such as the Duchess of Beaufort, frequently supported important horticultural innovations by helping to acclimatise plants imported from overseas.[14] For the most part, though, landscape gardening remained a purely aesthetic activity.

Plants were not only politically, economically, scientifically and aesthetically controlled in the eighteenth and nineteenth centuries; they were also physically contained in several different ways. Their transportation from the tropics and their cultivation in botanical gardens, commercial nurseries, the private estates of the wealthy and aristocratic, and eventually the conservatories attached to middle-class homes, required them to be enclosed. Whether in cases, orangeries, glasshouses, conservatories, winter gardens or palm houses, plants and flowers experienced levels of containment that they had not known in their natural state.

Up until the mid-nineteenth century a variety of containers were used to transport plants. Some were barrel-shaped, incorporating metal mesh to allow air to reach the plants; others were wooden-lidded boxes with compartments; others were open-ended, once again to let air in; and yet others were made of glass. It was not until 1829 that a breakthrough came, however, with the accidental discovery by a physician and amateur naturalist, Nathaniel Bagshaw Ward, of a plant transportation box that was

Fig. 4 A Wardian case, *c*.1850s. Wood and glass, 12.2 × 15.2 in.
Cases such as this were used to transport live plants in a sealed environment

left closed during journeys. It became known as the 'Wardian case' (fig. 4). The fact that it was sealed allowed water vapour to condense and run down inside its glass walls to be reabsorbed by the plants. In the ecosystem thus created, plants survived exceptionally well. The botanist, garden designer and author John Loudon was the first to realise the advantages of the new discovery, and the new cases were first used in 1848 by the botanist and plant hunter Robert Fortune for transporting tea plants to India from China. Their widespread use on ships from the 1850s onwards resulted in a huge growth in the importation to Europe of exotic plants from the tropics. As a result, commercial nurseries were able to stock many more such plants than hitherto, and to sell them at lower prices than ever before.

On their arrival in Europe, the exotic plants and flowers needed to be contained within controlled environments so as to ensure their survival. Besides those destined for commercial nurseries, by the eighteenth century many of the plants brought back by the travelling physicians, horticulturists, botanists and hunter-explorers were deposited in botanical gardens. These had developed from the mid-sixteenth-century physic gardens attached to several Italian universities, including those in Pisa, Florence, Padua and Bologna, which had focused on the academic study of medicinal plants.[15] The Italian model had spread to Europe through the

Fig. 5 The Temperate House at the Royal Botanic Garden, Edinburgh, built in 1858. Photograph by the author, 2016

sixteenth and seventeenth centuries, with new botanical gardens emerging in Spain and northern Europe. By the first half of the eighteenth century, yet more had been established in the North American colonies.

By the end of that century the Royal Botanic Gardens at Kew, originally established in 1759, and the Real Jardín Botánico de Madrid, which opened four years earlier, had moved from academic work to supporting international trade in plants. Under the informal directorship of Sir Joseph Banks, Kew, for example, sent collectors to the South African Cape, Australia, Chile, China, Ceylon and Brazil, among other places. New gardens were also established in the tropics to enable plants to be looked after locally. The British opened the Calcutta Botanic Garden in 1786; the French set up Pamplemousses Botanic Garden in Mauritius in 1735; and the Real Jardín Botánico de Madrid established the botanical gardens of La Orotava in Tenerife. By the early nineteenth century many more botanical gardens had been founded in countries across the globe, including Singapore, Australia, New Zealand, Ceylon, Egypt, South Africa and Russia.

Plant-filled greenhouses and conservatories were constructed in many of the botanical gardens. Built to support plants arriving from warm climates, one of the first appeared in Chelsea's Physic Garden in 1680. By the middle of the nineteenth century their numbers had increased significantly. The

Tropical Palm House in Edinburgh's Royal Botanic Garden was constructed in 1834; the Jardin des Plantes in Paris included a Mexican Hothouse, built between 1834 and 1836 by Charles Rohault de Fleury; and the palm house in Belfast's botanical gardens was constructed in 1839.[16] Many more emerged in subsequent decades. Later examples include Edinburgh's Temperate House, constructed in 1858 (fig. 5), and palm houses built in Copenhagen and Florence c.1874. Strasbourg's was built between 1877 and 1882; in 1907 the Tropical House in the Royal Botanic Garden, Dahlem, Berlin, was opened, and the Palm House in Amsterdam's Hortus Botanicus was completed five years later.[17] Similar developments took place across the Atlantic: the conservatory in New York's Botanical Garden, for instance, was built between 1899 and 1902.

It is difficult to pinpoint the exact moment at which the botanical garden conservatory changed from being a focus for the medical and scientific professions and horticulturists to becoming a destination at which people could spend their leisure time. By the late to mid-nineteenth century, however, that transformation was complete. Although several early botanical gardens – such as the one in Leyden – had been sites of education from the outset, they extended that service to the public only later. The Royal Botanic Gardens at Kew gave the public access in 1841, seven years before its Palm House, designed by Decimus Burton and Richard Turner, was complete, and twenty-two years before its Temperate House, also designed by Decimus Burton, was opened.

Once the conservatories in botanical gardens became destinations for public leisure, the way in which plants and flowers were arranged inside them changed. Abandoning the botanists' systematic approach, displays began to reflect the picturesque preoccupations of the eighteenth-century landscape gardeners. To add to the entertainment and level of spectacle experienced during a visit to such a conservatory, the creators of the displays sought to recreate a sense of the exotic atmosphere of the jungles in which many of the plants had originated. The fear of the jungle was evocatively described by Jean-Baptiste-Christophe Fusée Aublet, the French king's botanist. 'You have to be in a jungle to understand how dangerous it is to enter', he explained, adding that, 'The thorny trees, the tangle of razor sharp plants . . . leave the traveller fearing for his life.'[18]

By the second half of the nineteenth century, as well as visiting glasshouses in botanical gardens and commercial nurseries, people living in many towns and cities across the globe could also spend their leisure time in the numerous conservatories and palm houses that had sprung up in public parks. Funded either privately or by local municipalities, they provided a taste of exoticism that many visitors would not have been able to savour at home. In Britain the public parks movement was a key aspect of urban development at that time, having been established to provide free public leisure spaces for the working-class populations of places adversely affected by industrialisation. In the words of Frank Clark:

Fig. 6 Period postcard showing the palm house at Sefton Park, Liverpool, built in 1896

parks symbolised nature, affluence and health. Public parks were the answer to most of the ills of the time because of the strong belief of men like Francis Place . . . that a country-like environment transplanted to the city 'can always instil a hallowed calm, and a spirit of reverence into the mind and heart of man'.[19]

In 1896, a palm house was constructed in Sefton Park, Liverpool (fig. 6). It was stocked with a large collection of palms and other exotic plants, and was surrounded by sculptures by the French artist Léon-Joseph Chavalliaud, including one of Linnaeus (1900). Two sculptures by Benjamin Edward Spence, named *Highland Mary* (1854) and *The Angel's Whisper* (1857), were placed inside the building. Set in a bed of white lilies, the latter reflected the religiosity that could also be found in many of Liverpool's private parlours. The Isla Gladstone Conservatory was built in the same city in 1899, in Stanley Park, which had opened twenty-nine years earlier. It too was stocked with a large collection of exotic plants, including palms.

Late nineteenth-century London saw a chrysanthemum house constructed in Battersea Park, while Hackney's Victoria Park featured a palm house. Numerous European countries, as well as several other nations – South Africa, Australia and New Zealand among them – also saw palm houses and conservatories appearing in many of their public parks. In North America, likewise, buildings for plants were created in many of the new public parks. The wooden Conservatory of Flowers constructed in San Francisco's Golden Gate Park in 1879 was among the first (fig. 7). New York's Central Park, which opened to the public in 1859, also featured a

Fig. 7 The Conservatory of Flowers in Golden Gate Park, San Francisco, built in 1879. Photograph by the author, 2016

conservatory, constructed in 1898, while others were opened in Chicago's Jackson and Lincoln parks in the following decade. The populations of many other North American cities – such as Pittsburgh, Detroit and Cincinnati – could also enjoy visits to the palm houses in their local parks.

Orangeries, frequently constructed in the neoclassical style, had first appeared in Renaissance Italy and had quickly spread through mainland Europe. They were the first structures to be built in the private gardens of the wealthy to support the growth of exotic fruits – oranges, pomegranates and myrtles among them.[20] In Britain, the little orangery at Osterley House in London was designed by Robert Adam in around 1780, while Joseph Paxton created one at Hampton Court in 1846, five years before the construction of his Crystal Palace in Hyde Park for the 'Great Exhibition of the Works of Industry of All Nations'. In Germany, an orangery was built at King Frederick the Great of Prussia's summer palace of Sanssouci, in Potsdam, in 1864 (fig. 8).

Orangeries were soon joined by hothouses, which utilised greater quantities of glass. Some orangeries could be transformed into glasshouses by simply replacing their slate or tile roofs with glass.[21] Examples of very early domestic glasshouses include one made of timber and glass – more a cold frame than a greenhouse – designed by John Evelyn and illustrated in the 1691 edition of *Kalendarium Hortens*.[22] In 1689 a small glass case had been made for Queen Anne at Hampton Court, built by a Dutchman.[23] Over the

Fig. 8 The orangery at Sanssouci, the summer palace of King Frederick the Great, in Potsdam, built in 1864. Photograph by the author, 2014

following century the greenhouse, as it came to be called, was to experience a number of important, technology-led changes in relation to both its heating mechanisms and its materials. In 1817 John Loudon declared that a garden was not considered complete without a greenhouse or conservatory.[24]

By the eighteenth century many wealthy landowners had constructed greenhouses in their private walled gardens. Gradually, however, greenhouses were integrated into flower gardens and came within sight of the house. The trend for the garden to be seen as an extension of the house was much promoted by the landscape gardener Humphry Repton. Unlike his predecessor Lancelot 'Capability' Brown, whose landscapes consisted of rolling lawns that drew the eye away from the house, Repton reintroduced formal terraces, balustrades, verandas and trellis-work in order to bring house and garden back together.[25]

Impressive garden conservatories were constructed in many British locations, including the grounds of Syon House, outside London, and those of Bretton Hall in West Yorkshire, both 1827. The latter was one of the first conservatories built in the curvilinear style, made possible by Loudon's invention in 1816 of a flexible wrought-iron glazing bar. Examples were also created (by Joseph Paxton) in the grounds of Chatsworth House, the seat of the Duke of Devonshire, between 1836 and 1840, and, in the same decade, at Dalkeith Park, near Edinburgh. Later in the century an Italianate curvilinear conservatory was constructed in Ramsgate in Kent (fig. 9).

Fig. 9 A large flowering agave inside the curvilinear conservatory, King George VI Memorial Park, Ramsgate, Kent, 19th century. Photograph by the author, 2017

Soon it became the norm for conservatories to be attached to houses, so that the ladies who gardened in them could avoid inclement weather. As the advice writer Shirley Hibberd explained in the 1895 edition of his book *Rustic Adornments for Homes of Taste*, 'We will suppose a lady to be interested in plants, and the happy possessor of a well-furnished conservatory. If it is far removed from the dwelling-house, what a deprivation of accustomed amusement will a week of snow or frost occasion!'[26] In 1793, at Bellevue House in Ireland, a series of large greenhouses comprising a 650-foot-long conservatory, an orangery, a cherry house, a peach house and a vinery were attached to the main building so that Mrs Latouche, the lady of the house, could walk straight into them without risking an encounter with the elements.[27] At Orton Hall in Peterborough an attached conservatory in the Tudoresque style was constructed in 1835, while a fully integrated conservatory, filled with sculpture, fountains and pillars, was added to Somerleyton Hall in Suffolk in 1843.

Across the Atlantic in Montreal, the Shaughnessy House, constructed in 1875, contains a highly elaborate conservatory that features a patterned marble floor, engraved glass and ornamented iron pillars (fig. 10), while in the United States the fashion for extravagant domestic conservatories became a characteristic of the large houses built for wealthy members of the Gilded Age (c.1870–c.1900). Examples include William H. Vanderbilt's

Fig. 10 Hibiscuses flowering in the window of the ornate conservatory, Shaughnessy House, Montreal, built in 1875. Photograph by the author, 2015

mansion at 640 Fifth Avenue in New York, completed in 1882; the Louis Stern-Hugo Reisinger home at 993 5th Avenue in the same city, built in 1887; and Beacon Hill House of 1909, the Newport cottage of the railroad magnate Arthur Curtiss James, in the conservatory of which Mrs James spent a great deal of time reading about flowers.[28] A very particular example of American indoor landscaping could be found in Boston, in the Venetian-style museum and home of the art collector Isabella Stewart Gardner, who in 1903 created a glass-covered courtyard garden filled with palms and other tropical plants in the very heart of her building.

While attached conservatories were usually linked to drawing rooms – women's spaces – they were occasionally attached to morning rooms as well (fig. 11). Unusually, at The Grange in Hampshire, a conservatory was positioned opposite the lady's chambers. Both Repton and Loudon had strong views about attaching conservatories to houses. In 1816 Repton, who designed several conservatories, some with John Nash, advocated attaching them to the house in order 'to alleviate the parlour's formal gloom'.[29] He also advised the addition of a small aviary, or a flower passage, between the conservatory and the drawing room to avoid smells of stale earth and rotting plants coming into the house.[30] Some years later Loudon explained that owning an attached conservatory was 'not indeed one of the first necessities but one which is felt to be appropriate and which mankind

Fig. 11 'Conservatory Leading from Dwelling-House', an illustration from Shirley Hibberd, *Rustic Adornments for Homes of Taste*, new and rev. edn (London: W. H. & L. Collingridge, 1895), p. 136

Fig. 12 Photograph showing a British middle-class family sitting in a conservatory attached to the house, c.1910

recognises as a mark of elegant and refined enjoyment', adding that the smell of flowers wafting into the house was a definite benefit.[31] Loudon believed that women were naturally indoor people, and that the addition of an attached conservatory gave them and their daughters an engaging winter pastime.

The development of sheet glass and the repeal of the tax on glass in 1845 allowed conservatories to find their way from wealthy rural homes into urban middle-class houses. The movement began in Britain (fig. 12), where Paxton's Crystal Palace acted as a model for smaller domestic versions. When the fashion for conservatories eventually reached the United States, the American advice author Henry T. Williams explained in 1872 that, 'As an ornament of architectural value, no villa is complete without them', adding, 'We find nearly all the best styles and designs . . . offered only by English horticultural manufacturers.'[32] In 1876 Paxton launched a conservatory kit advertised as 'glasshouses for the million', thereby helping to extend the market for domestic conservatories even further.

The domestication, and accompanying feminisation, of the conservatory attached to the home bestowed specific meanings on the plants within them. As had happened earlier with landscaped gardens, they were transformed from objects of scientific interest and inquiry into cultural artefacts with considerable aesthetic and symbolic value. Exotic plants

Taming the Jungle

Fig. 13 Palms and ferns inside the octagonal conservatory at Palmyra, the house of Samuel J. Waring in Aigburth Vale, Liverpool, 1896

and flowers originating mostly in Australia, Africa and Asia were preferred. The much-favoured Cape heaths of the early nineteenth century were replaced a little later by cacti and orchids. The last became very fashionable and could be found suspended from the ceilings of conservatories in baskets. 'Once a luxury', Andrea Wulf has explained, they were now 'a sought-after accessory for the middle classes'.[33] Simultaneously, conservatories became less workspaces, and more places for leisure. Facilitated by the fact that plants were gradually removed from pots and planted directly into soil, gardeners began to emulate the practices followed in botanical gardens, and encouraged the development of picturesque interior landscapes. As Tovah Martin has evocatively expressed it, 'the wild beasts of horticulture were domesticated. Tropicals were tamed, potted and exhibited in idyllically tidy "jungle" scenes.'[34] Hothouse technologies continued to develop, such that by the early nineteenth century steam was being used to heat domestic conservatories. The smoke from braziers disappeared, leaving a clearer, moist atmosphere that offered a much more pleasant experience.[35]

Gradually, the word 'conservatory', in contrast to 'glasshouse', 'hothouse' and 'greenhouse', came to denote a space that not only was associated with plants, but also supported social interaction. As Henry T. Williams explained, the conservatory 'is mainly devoted to ornamental purposes, and the exhibition of plants in full beauty of growth and bloom, while in the humbler greenhouse, propagating boxes are the chief furniture used by the gardener for the production and forcing of his young plants', subtly implying that the concepts of greenhouse and conservatory had gendered connotations attached to them.[36] Other writers from the period encouraged the addition of fish, sculpture and seats in domestic conservatories in order to create congenial environments in which afternoon tea could be enjoyed and proposals of marriage made.

A striking example of a conservatory that acted as an extension of a home could be found in Liverpool, one of Britain's most important ports, into which numerous plant-filled merchant ships arrived regularly. In 1896 the furnisher and decorator Samuel J. Waring, who was to become a partner in the British furnishing and decorating firm Waring and Gillow, bought a house in Aigburth Vale, which he called Palmyra, the name of both a type of palm and the ancient site in Syria. Waring added an octagonal-roofed conservatory, which led to a series of elaborate, intersecting glasshouses attached to the rear of his early nineteenth-century villa (fig. 13). He filled the first with exotic plants, mostly palms but also a few ferns, and added some garden furniture and a standard lamp, thus transforming it into a semi-outdoor parlour. Chinese paper lanterns hung from one of the metal beams.

As the following chapter will demonstrate, in the private sphere plants and flowers next moved from the attached conservatory into the house itself, thereby securing their integration into the very heart of Victorian middle-class domesticity. Chapter 3 will then trace the legacy of the public sphere conservatory in the emergence in the late nineteenth century of a new concept – the commercial winter garden – which, while retaining its links to horticulture, became primarily an indoor pleasure garden for the new leisure classes.

CHAPTER 2

The Jungle in the Parlour

[T]he culture of choice plants in the greenhouse and the window, seem[s] to me more remunerative, both intellectually and morally, than even the study of the higher departments of art, because of their suitability to all tastes, and means, and their directly educative power, for they keep us near to nature and compel us to be students of the out-door world, whence many noble inspirations and devotional impulses are drawn.[1]
Shirley Hibberd

Shirley Hibberd's words were written in 1856, at a moment when tamed nature – in the form of plants and flowers, not to mention fish in aquaria, caged birds and snakes in terraria – was moving from the attached conservatory into the urban middle-class home. Given that flowers died and plants demanded nurturing, the substantial commitment that this area of domestic life required from the nineteenth-century housewife meant that some form of compensation was clearly necessary. Hibberd and others fully understood the extent of the repayment that could be expected, and articulated it in their advice books.

Plants had first made an appearance in the modern home in the Netherlands in the seventeenth century. The Dutch obsession with tulips, both as growing bulbs and as cut flowers displayed in *tulipières*, has been described as 'tulipomania'.[2] Markers of wealth and social status, tulips had been so central to the Dutch economy that the tulip market's eventual collapse brought about a financial crash in the Netherlands in 1637. By the late eighteenth and early nineteenth centuries plants had begun to enter the homes of the wealthy in Britain. At Stourhead House in Wiltshire, for instance, Thomas Chippendale Jr. created an indoor planter in the shape of an Egyptian sarcophagus for Richard Colt Hoare to use in his library.[3]

Fig. 14 'Ornamental Fern-Case Outside Window', an illustration from John R. Mollinson, *The New Practical Window Gardener*, new edn (London: Henry J. Drane, 1894), p. 23

Fig. 15 'A Warrington Case with Filmy Ferns,' an illustration from John R. Mollinson, *The New Practical Window Gardener*, new edn (London: Henry J. Drane, 1894), p. 97. The case combines an ornamental plant case with an aquarium

By the mid-nineteenth century industrialisation had made available an increasing number of furnishing and decorative items with which women could transform their middle-class homes into havens, safely separated from the outside world in which most remunerated work was undertaken by men. Much has been written about the subtle language of the interiors of the nineteenth-century urban and suburban middle-class home, emphasising the way in which its décor expressed its inhabitants' adherence to the fast-moving fashions of the day.[4] Plants and flowers played an important part within that. Although plucked from the natural world, they rapidly became (near) cultural artefacts in the domestic context, extensions of the furniture and furnishings that surrounded them, and significant markers of aesthetic knowledge, or taste.

Plants and flowers softened architectural frames, lightened heavy furniture, provided decoration, elegance and refinement, acted as screens (ivy, known as the poor man's vine, was particularly important in this regard), and added colour, texture and scent. The plants that best survived

Fig. 16 Ludvig August Smith, *Interior with Mother and Daughter by a Window*, 1853. Oil on canvas, 18.7 × 16.1 in. Nationalmuseum, Stockholm

gas lighting and coal fires included aspidistras, which needed little light or space, dracaenas, described as 'a favorite with all fond of the plant decoration of rooms', rubber plants, some palms, robust ferns and cacti.[5] The possibility of including plants and flowers in decorative schemes varied according to one's wealth and access to advice. If a family was unable to afford even a small conservatory, other ways of displaying plants and flowers inside the home were available to them. They included the use of enclosed boxes hung on to the exterior of windows (fig. 14), hanging baskets in porches and window bays or on balconies and verandas, fern cases (sometimes combined with aquaria; fig. 15), small pots on windowsills and mantelpieces or in cosy corners, ivy trailing over furniture items and walls (fig. 16), plants in pots on stands, and ferns displayed in fireplaces during the summer (fig. 17). Many versions of the Wardian case were developed for the home, while a wide range of other accessories, among them free-standing pots, plant stands, *jardinières*, *etagères*, wooden troughs with lead liners, and plant cabinets also facilitated the introduction of

Fig. 17 Ferns displayed in the fireplace of the drawing room at Terceira House, Aigburth Drive, Liverpool, 1891. Photograph by H. Bedford Lemere

Fig. 18 'Home-Made Wardian Case', an illustration from John R. Mollinson, *The New Practical Window Gardener*, new edn (London: Henry J. Drane, 1894), p. 126

nature into the domestic arena (fig. 18). The inhabitants of working-class homes often placed geraniums in pots on their windowsills, one of the very simplest and least expensive forms of window gardening.

While the most obvious use of domesticated plants and flowers in the nineteenth-century middle-class home was as visual components within interior décor, they also performed many other functions in that context, both practical and symbolic. The natural items displayed in nineteenth-century homes helped introduce children to the laws of science and post-Enlightenment rational thought, as well as playing a part in the self-improvement of adults. Before the popularisation of Charles Darwin's ideas, perhaps the most important symbolic meaning of domesticated plants and flowers, however, derived from the widespread belief that nature was made by God himself and was therefore inherently 'good'. As God was believed to be the creator of all things 'bright and beautiful', the beauty and goodness of indoor plants and flowers were seen as indistinguishable.

In their wild state, indigenous plants and flowers were available to everyone. Ferns could be acquired in the woods, and geraniums were easily propagated. The presence of nature in the home was also ennobling, as possessing plants, especially the non-indigenous, exotic varieties, had hitherto been the preserve of the wealthy and the aristocratic. Very importantly, domesticated nature was also seen as therapeutic, both in the sense that many people needed to negotiate new relationships with both the pre-industrial past and the modern industrial present and future, to which they were adapting often with some difficulty, and because living nature could act as a companion to the lonely. The anxiety and fear excited by raw nature's power was, in the domestic context, replaced by

Fig. 19 A palm and other indoor plants and flowers ornament an Edwardian dining room, 1905

a belief that nature had a calming effect on the soul. Hibberd believed that nature inside was a source 'of rest, and solace, and refreshment'.[6] Henry T. Williams, who was of the same mind, explained that indoor flowers and plants were 'soul-refreshing, heart-inspiring and eye-brightening'.[7] He went on to claim that plants could calm children at school, amuse invalids and even provide an object of affection for bereaved parents.[8] The Victorians understood the comfort that came from being reunited with nature in its tamed, domesticated form, and believed there to be a close, emotionally charged relationship between humans and plants based on their both being living organisms.

In its non-indigenous form, nature inside was also a stimulant for exotic fantasy and an escape from the domestic chores of everyday life. In addition, it represented the values of an era that was still predicated upon colonial expansion and the need for a strong sense of national identity. Plants and flowers were also believed to perform a number of utilitarian roles in the home, including producing useful gases. Anticipating ideas that were to become widespread in the early twenty-first century, G. Lister Sutcliffe explained in his 1898 book *The Principles and Practice of Modern House Construction* that 'They ... preserve the purity of the air

Fig. 20 Palms in the dining room at Palmyra, the house of Samuel J. Waring in Aigburth Vale, Liverpool, 1896. Photograph by H. Bedford Lemere

by removing the poisonous gas evolved by animals and the combustion of hydrocarbons and maintain the equilibrium of nature'.[9]

Given the wide range of their possible meanings and roles, plants and flowers in nineteenth-century British, continental European and North American domestic interiors were neither frivolous nor superficial decorative afterthoughts. Nor was their introduction merely a way of complementing the bric-a-brac in Victorian parlours, or of avoiding the *horror vacui* that so preoccupied nineteenth-century housewives. Rather, placing a dracaena in a ceramic pot on a stand in the bay window of a Victorian parlour, putting a potted Kentia palm in the corner of a dining room filled with dark wooden furniture, arranging ferns in a fireplace for the summer, trailing ivy around a door frame, or positioning a basket of chrysanthemums on the dining table constituted highly significant interventions which, whether their instigators knew it or not, touched the aesthetic, social, psychological, spiritual, moral, cultural, economic, political, scientific and technological nerves of daily life.

Of all the household plants of the era, the palm was undoubtedly the most exotic. Seen as tasteful, mysterious, tropical and graceful, palms enhanced countless parlours and hallways (fig. 19). They fulfilled

The Jungle in the Parlour

Fig. 21 'A Modification of the India Table Decoration', an illustration from Shirley Hibberd, *Rustic Adornments for Homes of Taste*, new and rev. edn (London: W. H. & L. Collingridge, 1895), p. 19

an important compositional role, frequently providing a frame – a set of stage side curtains, as it were – for the other interior elements. They could also act as screens, and frequently served to unify eclectic, cluttered settings. They also added height where it was needed, and the colour green to complement the widespread use of deep reds that pervaded the Victorian parlour. Many of Britain's domestic palms were supplied by the Loddiges family's Hackney Botanic Nursery, which contained the world's largest hothouse.

As the century progressed, palms became increasingly visually integrated into fashionable interiors, both in wealthy country houses and in more modest urban middle-class dwellings. They introduced the exoticism of the tropics, as well as memories of empire and of an untamed world in which nature had held sway over culture. Samuel J. Waring, whose attached conservatory at his Liverpool home, Palmyra, was mentioned in Chapter 1, did not limit his interest in palms to the conservatory. He also introduced some into his dining room (fig. 20). The room in question was highly conventional in its eclecticism and historicism. It featured Chippendale-style dining chairs, French-eighteenth-century-style electric sconces, a Japanese-style bamboo fire-screen, and a neo-Tudor moulded

Fig. 22 Ferns and other indoor plants surround an elderly middle-class couple photographed in an Edwardian parlour, 1910

plaster ceiling. The Turkish or Persian carpet on the floor was a sign of the fashion for orientalism that characterised many middle-class domestic interiors of the period. The addition of those three small plants, one of them positioned at the centre of the dining table, contributed to that taste of exoticism.

Hibberd's *Rustic Adornments for Homes of Taste* provided a description of an 'India table decoration', the central element of which was 'a handsome feathery palm' (fig. 21).[10] Tables with holes at the centre through which protruded growing palms, with their pots on the floor, also graced many dining rooms. The aim was to produce an elegant outline without obstructing the view of the guests. Several species could survive in domestic settings, among them the Kentia palm, originally from Norfolk Island in the South Pacific, which had been colonised by Captain James Cook in 1774.[11] In 1869 Sir Joseph Hooker, the curator of Kew's Royal Botanic Gardens, was sent some Kentia palm seeds by Charles Moore, the director of the Royal Botanic Garden in Sydney. It quickly became a favourite in wealthy British settings due to its elegance and its ability to flourish indoors.[12]

Ferns, as well as being introduced to Britain from Jamaica, New Zealand, India, Mexico and Japan, could also be found in local woods and, though

The Jungle in the Parlour 41

Fig. 23 A lady tends the plants in her window box, 1878. Engraving by an unknown artist in *Cassell's Family Magazine* 4, no. 9 (August): 564

they were less robust, they offered a very cheap and accessible alternative to the potted palm.[13] Like palms they had a soft profile that offset the hard woods of interiors, and the strong green of their foliage once again complemented the palette used in Victorian parlours (fig. 22). Their popularity increased with the development of a reliable method for raising them from spores.[14] The phenomenon named 'pteridomania', or fern madness, which in Britain lasted until the end of the 1860s, has been extensively documented.[15] In Britain fern-hunting was undertaken in the woods at night, while ferns could also be purchased from the Loddiges family's Hackney nursery. As well as providing elegant parlour decorations they were frequently pressed, framed and mounted onto parlour walls.

In both Britain and the United States, engaging in indoor plant- and flower-related activities was considered a natural extension of nineteenth-century middle-class women's participation in outdoor gardening. Horticulture and agriculture had traditionally been considered highly specialised masculine domains but, from the 1840s onwards, a close link developed between middle-class women and gardening in the domestic context. Jane Loudon's *Instructions in Gardening for Ladies* was published in Britain in 1840, and was soon followed by other works.[16] Loudon's books signalled a new approach that encouraged the engagement of female novices.[17] Even so, some readers seem to have found her texts inaccessible. Across the Atlantic, in the 1844 edition of her influential

Fig. 24 The colourful front cover of Webbs' *Bulb Catalogue*, 1888

Fig. 25 'Arrangement of Plants and Hanging Baskets in Window', an illustration from John R. Mollinson, *The New Practical Window Gardener*, new edn (London: Henry J. Drane, 1894), p. 24

book *Every Lady Her Own Flower Gardener* (first published in 1839), Louisa Johnson noted that her companions had asked her to write 'without Latin words and technical terms' such as Loudon employed, so that amateurs like themselves could learn how to become gardeners.

The idea that window gardening was an amateur artistic practice undertaken by women emerged in the context of the distinction between paid work undertaken outside the home, which was seen as masculine and rational, and home-making, which was unpaid and defined as feminine, emotional, moral and, above all, aesthetic (figs 23 and 24). Women's domestic creative work in the nineteenth century, which comprised activities often referred to as 'feminine accomplishments' (for example, shell-work, hair-work, leather-work, work with moss and pine cones, Berlin wool-work, crochet and embroidery, as well as arranging plants and flowers, making floral wreaths, bouquets and table decorations, and pressing and mounting ferns and flowers), grew apace in the period, supporting the never-ending quest for domestic beauty and the social status that beauty was seen to mark.[18]

Many of the symbolic meanings communicated by indoor plants and flowers, including their gendered implications, were outlined in the numerous window- (British) or parlor- (American) gardening advice books that were written between 1820 and 1900. A body of writers on both sides of the Atlantic set out to instil the necessary knowledge and skills in those new to the art of introducing plants and flowers into their interiors, and to explain the benefits to them (fig. 25). As the commercial market for such books was rapidly expanding, in the years after 1840 publishers commissioned them in increasing numbers and many new editions were printed.[19] European-language texts were translated into English, and many English-language books were published on both sides of the Atlantic, often adapted appropriately for the different markets. Several American books used the same images as had been used in British publications, suggesting that publishers were collaborating closely in the expanding market and that many of the ideas were easily transferable.

As with all areas of advice literature, it is hard to know whether the guidance offered was actually heeded; nonetheless, the books provide an insight into the possibilities that were seen to exist.[20] They covered the whole spectrum of window gardening, from planting and growing plants and flowers for indoor use, to maintaining them once inside, making and acquiring a wide range of material artefacts in which to display them, and arranging plants and cut flowers as a component of interior décor. Much advice was also provided in the numerous popular gardening and home journals that came into being in those years.[21] Advice books on both sides of the Atlantic broadly agreed that plants and flowers in the home represented an important form of compensation for the loss of nature experienced by the newly urbanised population. One of their most significant messages related to the assumed roles of men and women.

In general terms, those roles shifted as the century progressed, from a fairly equal division of labour at first – men doing the practical and horticultural activities and women undertaking flower arranging and introducing plants and flowers into their interior decorative schemes – to a situation that saw women taking on a greater share of the work.

In Britain, early advice on how to look after plants in pots was provided by Elizabeth Kent and Leigh Hunt in their 1823 book *Flora Domestica; or, the Portable Flower Garden*, in which the authors showed an early awareness of the tension between the scientific field of botany and the art of arranging plants and flowers. Thirty-three years later, Hibberd claimed that, in the form of domesticated plants and flowers, the world of nature provided a metaphor for the highest achievements of human nature itself and offset the 'commotion and dust' that otherwise dominated life in an urban setting.[22] His *Rustic Adornments for Homes of Taste* (first published in 1856 but reprinted in 1857, 1870 and 1895) was among the first British books to address the subject of window gardening that reached a large audience.[23] Its main emphasis was on the construction of what the author called 'toys' for the home, fish tanks, bird-houses, bee-houses and plant cases among them. Most of the 1856 and 1857 texts dealt with aquaria, aviaries, apiaries and the exterior garden, while only three small sections of the book, entitled 'The Wardian Case', 'The Waltonian Case' and 'Floral Ornaments for the Table and Window', were dedicated to the inclusion of plants and flowers in interiors. While the first two of these – clearly directed at a male reader – detailed the construction of different kinds of cases and the technicalities of growing plants in them, the third addressed a female audience. Readers were therefore led to believe that, in the 1850s, window gardening involved a gendered division of labour.

In his introduction, 'The Home of Taste', Hibberd ensured that his female readers did not feel excluded by acknowledging that he believed indoor gardening to be an art form. He spoke of the 'union of Nature and Art', and in the preface he used the terms 'domestic aesthetics' and 'domestic elegance', which he believed to have existed in middle-class society for around twenty years.[24] Hibberd's emphasis on taste, and the link between window gardening and other artistic practices, singled him out from other writers on the subject. 'No matter in what form', he wrote, 'the cultivation of Taste may manifest itself, in paintings and sculptures, in the analysis of scenery, in the grouping of flowers, in the embellishment of the window or the mantel . . . refinement of manners, kindliness of feelings and a deeper devotion will be its sure attendants'.[25] Hibberd was undoubtedly conscious of the special relationship between women and plants and, when it came to the subject of decorating and arranging the latter, he addressed a female audience directly, exhorting his readers to 'adorn your table . . . with flowers out of the fields'.[26]

Across the Atlantic, Henry T. Williams's book *Window Gardening* was first published in 1862 in New York. Later editions followed in 1872 and

1876. It owed much to Hibberd's text, with the emphasis in the first two sections being on toys, the categorisation of plants, and the practicalities of growing plants indoors, including how to keep insects at bay. The concept of taste was mentioned only in passing, however, and ladies were included only briefly in their capacity as natural nurturers. Williams noted at one point that it was 'easy for many [a] lady to raise her own verbenas'.[27] He also included an engraved image of a conservatory featuring a female gardener and made a reference to Miss Maling – a well-known advice writer herself, as well as an inventor – crediting her with the design of a large fern case.[28] In the third section of the book, entitled 'Parlor Decorations', women were, predictably, addressed more overtly.[29] Williams openly acknowledged that women had been very successful in that area. 'The amateurs [i.e., women] have outstripped the professionals in the rapidity of their progress', he wrote, adding that 'the prettiest rooms today are embellished by the fingers of a fair plant lover, who a year ago did not know one flower from another'. There was an implication in Williams's text that parlour decorating required more engagement and activity than many other forms of women's creative work in the home, presumably because some knowledge of floriculture was required.

Two books relating exclusively to the subject of cut-flower arranging were published in 1862. One was written by Mr Thomas March, who had won first prize at the first Royal Horticultural Society show of 1861 for the design of what, predictably, came to be called a 'March stand'.[30] March's book of the following year, *Flower and Fruit Decoration*, which looked back to the era of the country house, told (male) gardeners which kinds of vases and soil to use. The target audience throughout his short text was the male tradesman who supplied the materials and flowers for table decorations in wealthy households. John Perkins – from 1848 the gardener to the Henniker family at Thornton Hall, Suffolk – adopted the same approach in his 1877 book, *Floral Decorations for the Table.* Like March, Perkins remained entrenched in an earlier era in which male gardeners had held sway, and largely ignored the way in which middle-class ladies were taking over the art of table decoration.[31]

Miss A. E. Maling's book *Flowers for Ornament and Decoration* was published in the same year as March's. The author acknowledged that the facet of indoor gardening she was describing was of interest to women only.[32] She covered a wide range of floral ornaments, from knots of flowers to wear on dresses, to bouquets, wreaths, and flowers in vases and hanging baskets. There can be no question about Maling's strong belief in the artistic status of flower arranging and in its being one of women's domestic accomplishments. As we have already seen, Maling was an inventor as well as a flower arranger and she won prizes for, among other designs, her invention of a fern case that could be either heated or cooled as required. In her determination to reach a female audience, Maling made frequent references to the links between flower arranging and dress. 'It is doubtless most desirable', she wrote, 'that flowers should suit the general colour and style of the dress and ribbons with which they are to be worn.'[33]

Annie Hassard, the author of *Floral Decorations for the Dwelling House: A Practical Guide for the Home Arrangement of Plants and Flowers*, published in 1875, extended Maling's approach to yet another generation of British middle-class female home-makers, and this time included plants as well as cut flowers. By this date amateur women were competing openly with professional men in the numerous flower-arranging competitions that were taking place around the country. Although overtly 'Intended for the use of Ladies and Amateurs', Hassard's book offers a very detailed account, both practically and artistically oriented, of the best plants and best pieces of equipment to use for a wide variety of indoor plant and flower decorations, from bouquets to dining tables, window displays, hanging baskets and Christmas decorations, as well as giving advice on how best to arrange them. The following year saw an American edition of Hassard's book published by Macmillan. Adapted for that market, this edition put even more emphasis on plants and flowers as part of the interior decorating schemes used in parlours and dining rooms.

In a later book, *The New Practical Window Gardener*, originally published in 1877, the British writer John R. Mollinson also focused on the construction of window gardening equipment, from window boxes to window greenhouses, with a male audience in mind. Towards the end of the book, however, his way of addressing the subject of floral decoration in the home was to let Annie Hassard speak on his behalf. In so doing, Mollinson was bowing to what he believed to be women's superior capacity for flower arranging. He was also showing that, while he was comfortable providing such practical tips as 'pot plants arranged in zinc pans and flower stands should always be provided with flats to stand in', he clearly felt less qualified to write about taste.[34] Like Williams before him, he openly admitted that 'Ladies with their nimble fingers and quick fancies are always the best at floral decorations in the dwelling. This is as it should be.'[35]

There was a tendency for female writers to focus more, albeit not exclusively, on flowers and flower arranging than on growing plants. That fitted with the prevailing idea that any self-respecting lady should know all the old-fashioned flowers and understand the language of flowers. By the 1880s, Japan was having a very strong influence on what had become a very specific art form.[36] As Charles F. Warner explained in 1910:

> If we have learnt much from the Japanese in regard to the arrangement and hanging of pictures, from them we have learnt more about the artistic arrangement of flowers. They have taught us to value the stem and leaves of the flower as essential to an artistic arrangement, that flowers of the same kind should be grouped together, and that harmony and blending of colour are necessary to secure the most artistic effects.[37]

The last years of the century saw the emergence of a male-dominated campaign against flowers. While flowers were seen as being overtly sexual, foliage was considered tasteful and chaste, and was said to call forth higher emotions.[38] A wedge was thus driven between plants and flowers that continued into the twentieth century, such that the cultures of household plants and flower arranging followed increasingly independent routes. While flower arranging remained within the realm of interior decoration, both amateur and professional, and was linked almost exclusively to female practitioners and the domestic sphere, household plants moved into the public arena, where the scale of operations became much larger and more overtly commercial.

By the turn of the century there was a strong sense that the values communicated by indoor plants and flowers in the early to mid-century, especially their links to science, education, spirituality, godliness and morality, had become overshadowed by their secular, aesthetic role in the fashionable domestic interior. Less and less was written about plants' religious and therapeutic functions. With that shift, nature inside became instrumentalised in the commercial arena. While the spread of indoor nature beyond the domestic sphere considerably eased the tensions associated with people's exit from the home and entrance into the world of commerce and public leisure, it also benefited the entrepreneurs, who were eager to take financial advantage of the benefits that nature inside had to offer.

CHAPTER 3

The Parlour Outside

Are plants sought, discovered, researched, propagated, bred, bought and sold for the benefit of plants or to satisfy the interests of people?[1]
Charles A. Lewis

Through the second half of the nineteenth century tamed plants and flowers were introduced into numerous new, urban, public and semi-public inside spaces across the Western world, many of them defined by their links to commerce. In those new settings, the traditional values and meanings that plants and flowers had communicated in the domestic sphere were joined by new ones linked to fashion, conspicuous leisure, pleasure, consumption and democratised luxury, in short, to the range of activities that defined the experience of urban modernity in the public sphere from 1850 to 1914.[2] The presence of plants and flowers in those spaces (or in some instances the mere memory of them) helped visitors feel sufficiently at home to want to engage in consumption.

In the public arena plants and flowers moved beyond the conservatories in botanical gardens and the palm houses in local parks into the inside spaces of urban and seaside winter gardens, exhibition buildings, leisure complexes and museums, and into the palm courts located within the semi-public inside spaces of grand hotels, ocean liners and department stores. From the seventeenth century onwards, glasshouses containing tropical plants had been looked upon as Gardens of Eden, or as little Paradises on earth.[3] Although now linked to commercial values, that vision continued to be associated with the new indoor public spaces of the nineteenth century.

Many public winter gardens were built in the second half of the nineteenth century. While they recalled the elaborate eighteenth-century orangeries and glasshouses constructed by the aristocracy on their extensive

Fig. 26 Phoebus Levin, *The Dancing Platform at Cremorne Gardens*, 1864.
Oil on canvas, 26 × 42 in. Museum of London

estates, and though they often contained plants, these were primarily social spaces. They did impart some knowledge of the natural world to their middle- and working-class visitors, many of whom were new to the experience of public leisure, but, more importantly, they helped transfer into the public arena the social interactions that had hitherto taken place in the domestic sphere. As a result, sitting, dining, socialising, parading and shopping in an indoor public environment containing exotic plants and flowers became a possibility for increasing numbers of people.

As leisure became democratised, the need for enclosed spaces in which public entertainment could take place increased. In eighteenth-century London the aristocracy had visited open-air pleasure gardens. Following the closure of the fashionable Vauxhall Gardens in 1859, Cremorne Gardens in Chelsea (fig. 26) became a very popular venue until its closure in 1877.[4] As well as lawns, groves and flower beds it contained a number of enclosed spaces, including kiosks, dance floors, refreshment rooms and its own iron-and-glass building, Ashburnham Pavilion, which was constructed in 1858.[5] The increasing need for covered leisure spaces converged with a growth in the supply of plants and flowers that could be used indoors.

Several public winter gardens combining horticultural and social functions had already appeared in continental Europe in the early nineteenth century. In his *Encyclopaedia of Gardening* of 1850 John Loudon described one example,

Fig. 27 Joseph Nash, *The Transept*, colour engraving published in *Dickinsons' Comprehensive Pictures of the Great Exhibition of 1851, from the Originals Painted for H.R.H. Prince Albert, by Messrs. Nash, Haghe, and Roberts, R.A.* (London: Dickinsons Brothers, 1854)

located in Berlin. Stefan Koppelkamm has explained that Loudon may have seen the Berlin winter garden on his first visit to Europe in 1813–14, when he also saw others in Strasbourg, Potsdam and Vienna.[6] Loudon described the Berlin building as 'stone-built with a solid roof and filled in winter with scented bulbs such as hyacinths and narcissi, plus heaths and acacias'.[7] According to Loudon it also contained a number of botanical novelties, including oranges and myrtles – the luxuries of the eighteenth century.[8]

Although, at that early date, the exotic plants in the Berlin winter garden were undoubtedly its primary attraction, other activities went on in it as well. According to Loudon, it was also a place for promenading and for seeing and being seen. He explained that 'It is almost needless to say that in these gardens there are plenty of seats and small moveable tables; there are also, generally, bands of music, a reciter of poetry, a reader, a lecturer or some other person or party to supply vocal or intellectual entertainment.'[9] Trunks of trees grew out of the middle of round tables, and billiard tables were provided for ladies, while the gentlemen read newspapers, took chocolate and talked politics.

The first conservatory in the Jardin d'Hiver in Paris was pulled down in 1846 and a new one opened two years later. As well as featuring 'promenades, a *jardin anglais* complete with a lawn and several fountains'

The Parlour Outside

and providing a feast for the senses, the 'air [being] filled with the scent of orange blossom and the song of birds', the new Parisian winter garden also contained an indoor ballroom, a café and a reading room.[10] However, as the appeal of indoor exotic plants was still new in France, the main focus remained on the garden.[11] Joseph Paxton attended a concert at the garden in its opening year and apparently admired the interior landscape he encountered there.

Britain's first large public winter garden opened in 1844, on land in Regent's Park which had previously been owned by Jenkin's nursery. It was designed by Decimus Burton and Richard Turner, and run by the Royal Botanic Society, which had been formed in 1839. The London winter garden was reserved for an educated public. As well as housing plants, it contained a lecture hall, a library and a museum, and it was used for lectures and meetings as well as for flower shows and social gatherings. Queen Victoria was the garden's first patroness and seems to have taken a great interest in it, encouraging ladies to become members.[12]

By the end of the nineteenth century, the term 'winter garden' had come to describe a spectrum of buildings: those containing plants, which could also be used for social purposes, and more hybrid leisure and entertainment complexes, which, while frequently containing plants, prioritised other functions. As the nature of urban social activities was changing at an unprecedented rate, and people were experiencing a traumatic break from much of what they had previously known, the presence of plants and flowers in those public leisure spaces maintained an important link to the past. The fact that they were phased out quite slowly minimised the shock that might otherwise have been felt. Indoor plants in public winter gardens played a transitional role, therefore, easing the path to the development of mass leisure spaces in urban settings.

The best-documented English example from the middle years of the nineteenth century is the Crystal Palace (fig. 27), which was designed by Paxton to house the Great Exhibition of 1851 in Hyde Park. It is often seen as one of the first international public buildings to be made from cast iron and glass, and was designed to act as a container for manufactured commodities and to engage the public in a modern experience that was a mix of education and entertainment. Paxton based the Hyde Park structure on his designs for greenhouses in the grounds of the home of the 6th Duke of Devonshire, Chatsworth House, between 1836 and 1840. Prince Albert and Henry Cole (a member of the Commission responsible for mounting the event) wanted the Great Exhibition to contain a display of manufactured products as a means of promoting international trade, and, in sponsoring the creation of the Palace, they set in train a trajectory that saw iron-and-glass buildings being used, later in the century and beyond, for exhibition halls worldwide.[13] The Crystal Palace was also strongly reminiscent of the horticultural glasshouse, however.[14] In addition, numerous elms were left on the site, a condition of the Palace's being constructed in Hyde Park, while

Fig. 28 Fountain and exotic plants in the north nave of the Crystal Palace, Sydenham, 1911

other trees were brought in to create a winter garden inside the building. When Paxton rebuilt his Crystal Palace in Sydenham in 1854 (fig. 28), the new iron-and-glass construction returned more closely to its origins as a glasshouse, housing, among other plants, a vast collection of exotics that he had bought from the Loddiges family's nursery when it closed in the 1850s.[15]

The Crystal Palace provided a model for a wide variety of subsequent iron-and-glass buildings, which included museums, exhibition halls, public leisure centres and department stores. While many arcades, marketplaces and railway stations also owed their design to the Crystal Palace, they rarely contained trees, plants or flowers. However, arguably, through their use of iron and glass they too evoked memories of the plant-filled glasshouse. Following the establishment of the original Sunderland Museum in 1846, a new one was built in 1879 next to Mowbray Park. Based on the Crystal Palace, it included a library and a winter garden with plants. Many others were built in other parts of the world. New York constructed its own Crystal Palace in 1853. In 1887 another example was built in the Jardines del Retiro in Madrid for the Exposition of the Philippines. Other large world

The Parlour Outside

Fig. 29 The interior of the Royal Aquarium and Summer and Winter Gardens, Westminster, London (opened 1876). Engraving by an unknown artist in the *Illustrated London News*, 16 October 1875

exhibitions housed in similar iron-and-glass buildings included the 1862 event held in London's South Kensington and the Vienna International Exhibition of 1873. The latter featured an enormous rotunda. The ambitious Centennial Exhibition held in Philadelphia in 1876 showed its exhibits in several separate buildings, most of which used iron and glass in their construction. The 1900 Paris Exposition Universelle included an iron-and-glass Horticultural Hall that was filled with plants and flowers.

A number of the new educational and leisure centres that emerged in London from 1870 onwards also contained winter gardens modelled on the Crystal Palace, among them the Alexandra Palace in Muswell Hill, which opened in 1873 but was destroyed by a fire and reopened in 1875; the Royal Aquarium and Summer and Winter Gardens in Westminster, opened in 1876; and the People's Palace on the Mile End Road, opened in 1886. These enclosed structures sheltered visitors as they performed their new social identities in public. Based on the conservatories in botanical gardens and public parks, they were run commercially. As a result, while the utopianism associated with the horticultural glasshouse persevered, the plants and flowers in those buildings also acted as commercial props.

The Royal Aquarium and Summer and Winter Gardens (fig. 29), a stone-and-brick entertainment complex, had a great hall covered with a barrel-vaulted roof of cast iron and glass, and contained palm trees, fountains, sculptures, thirteen large fish tanks and an orchestra.[16] Around the hall were rooms for eating, smoking, reading and playing chess, an art gallery, a skating rink and a theatre. A private enterprise – its original board of directors included the theatrical manager Wybrow Robertson and the retailer William Whitely – the Aquarium was intended to be a centrally located Crystal Palace. However, it was also meant to serve as an enclosed pleasure garden, replacing Cremorne Gardens.

Although, as its name implies, the Royal Aquarium and Summer and Winter Gardens was ostensibly dedicated to the display of fish, it also contained many trees and plants. In 1875 the secretary to the board, Bruce Phillips, had explained that, 'It has been found that visitors to other aquariums liked to have other amusements as well as observing the fish. The directors have therefore determined to have a botanical display and there would be a Summer and Winter Garden.'[17] Another early promoter of the building pointed out that 'The grace and freshness of a winter garden will be a great attraction, the hall surrounded by palms and exotic trees and shrubs, the whole having the general aspect of a vast conservatory filled with splendid sculpture.'[18]

The People's Palace, which combined the Beaux Arts building style with the materials of the Crystal Palace, was built to provide educational and leisure activities for the working-class population of the East End.[19] It contained, among other facilities, a winter garden, a concert hall, a reading room and a technical college. 'The People's Palace brought culture, entertainment and education to enrich the lives of the local people', explained one commentator, while *The Times* described the Mile End building as 'a happy experiment in practical Socialism'.[20] The winter garden, filled with palms, flowers and tropical fruit, was situated to the rear of the Palace's Queen's Hall. Begun in 1890, it was completed two years later and was used for concerts and refreshments (fig. 30).

As well as providing an education, both the Royal Aquarium and the People's Palace offered their visitors the possibility of developing good taste, and a variety of opportunities for healthy recreation. While the inclusion of plants in both venues provided an education in natural history, the very concept of the Aquarium was rooted in the idea that exposure to sea-life enhanced people's scientific knowledge of the subject. Indeed, aquaria – both domestic and public – multiplied from the 1850s onwards.[21] At the opening ceremony of the Royal Aquarium in 1876 Prince Alfred, Duke of Edinburgh, explained that 'The extensive aquarium, which is the main object of this institution, cannot fail, if properly directed, to stimulate the love of natural history and the acquirement of scientific knowledge.'[22] The People's Palace also housed a free library, opened in 1887, and a reading room, opened a year later. The library was open to the general public, as well as to students taking classes at the Technical Trade and Science Schools that launched in 1888. Classes in tailor's cutting, carpentry, photography, needlework, French and book-keeping were among those on offer.[23]

Fig. 30 Inside the winter garden at the People's Palace, Mile End Road, London (opened 1892), c.1892–1937

As the Duke of Edinburgh observed at the Aquarium's opening ceremony, 'The access to a useful reading-room, the daily performance of good music by a well-chosen orchestra, the periodical exhibition of such collections of paintings as we see around us – these are agencies which cannot but exercise a most beneficial influence in refining and cultivating the public taste.'[24] Sir Arthur Sullivan was in charge of the music (for a short period), and John Everett Millais took charge of picture selection. A theatre, opened in April 1876, remained in use until 1900, when it was taken over by Lillie Langtry, who commissioned a completely new theatre that moved out of the Aquarium. High-cultural activities were also pursued at the People's Palace. As well as the concerts which took place in the winter garden, organ recitals were given in the Grand Central Hall. Art lessons were also hugely popular.

In terms of physical activities, visitors to the Aquarium could engage in skating, cycling and boxing, while those who frequented the People's Palace could swim in the pool, exercise in the gym, roller-skate in the basement, play tennis and billiards, and dance. Traditional pastimes, including horticulture, were also encouraged at both spaces. Three shows of the National Chrysanthemum Society were held annually at the Aquarium, and similar displays also took place at the Palace. Much also went on in the Aquarium that was widely considered neither edifying nor health-giving, however, including the firing of a woman from a cannon, the exhibition of the first gorilla seen in England, and displays by performing animals and a scantily clad trapeze artist. Popular entertainment inevitably proved more lucrative than education and culture. Emphasis on such spectacles proved a short-term strategy, however, as the Aquarium's loss of reputation caused it to become desirable to nobody and it was forced to close in 1903.

Although the Palace did not sink quite so low, even there a tension emerged between the educational and entertainment aspects of its provision. The flower, poultry, pigeon, dog, cat and rabbit displays, which became regular events, were nevertheless relegated to sideshows, and were considered by some to be trivial pastimes compared to the more serious education provided by the classes, concerts and lectures. It was the entertainment that kept the books balanced, however. Like that of the Aquarium, the popularity of the Palace waned as the new century approached. A fire destroyed most of the building in the early 1930s, a new building was constructed, and the educational side eventually became part of Queen Mary College, University of London.

Although the Aquarium and the Palace served different purposes, they both contained plant-filled winter gardens. The inclusion of nature inside those spaces helped facilitate a seamless transition from the social activities of the nobility, undertaken in their private conservatories and public pleasure gardens, to mass urban leisure. In the end, however, the utopian idea of people maintaining contact with nature in an urban environment was undermined by the desire to make money and to appeal to a mass audience.

Fig. 31 Inside the conservatory at the winter garden in Southport, Merseyside, constructed in 1874

The result, arguably, was that after the turn of the century, the London-based winter gardens became less and less popular.

In contrast, the winter gardens at the British seaside lasted well into the middle of the twentieth century.[25] They too aimed to provide opportunities for leisure and entertainment, and contained displays of plants and flowers. The seaside locations first became popular with wealthy visitors in the eighteenth century. As the nineteenth century progressed, the middle classes and eventually the working classes began to holiday in them as well. The unpredictability of the British weather meant that visitors needed enclosed spaces in which to parade and be entertained. The plants within those spaces encouraged holiday-makers to relax and engage in various forms of consumption.

Advertised as the first and the largest of its kind, Southport's winter garden, located on England's northwest coast, was constructed in 1874. It combined an iron-and-glass conservatory with a second building that contained a concert hall (fig. 31). Like the Royal Aquarium, its early high-cultural ambitions proved unsuccessful financially and it ended up as a ballroom and roller-skating rink, while its concert hall was eventually transformed into a cinema. The managers of Blackpool's winter gardens were equally ambitious (fig. 32).[26] Opened in 1878, the site combined a

Fig. 32 Period postcard showing the interior of the Floral Hall at Blackpool's winter gardens, c.1910

concert hall and a theatre with areas for promenading, including a Floral Hall and a palm-filled winter garden. The intention was to 'convert the estate into a pleasant lounge, especially desirous during inclement days', thus demonstrating its essential domesticity.[27] A ballroom was added in 1896, together with an Indian Lounge.

The south coast of England was also popular with holiday-makers. In 1875 Bournemouth opened its winter gardens. On the east coast 1903 saw the opening of Great Yarmouth's equivalent (figs 33 and 34). Originally built in Torquay, Devon, twenty-five years earlier, it had been transferred to the east coast, to what was by then an archetypical working-class seaside town. As Darren Barker has explained, it was:

> Stuff full of exotic plants, a theatre of botany, which allowed the paying public the chance to see glimpses of faraway places, through an eclectic collection of plants from all corners of a flagging empire and beyond ... For the millions of holiday makers, packing the resort in the early decades of the twentieth century, escaping for a few days from the factories and the daily grind, the Winter Gardens was an unexpected paradise. As much part of the experience as the sticks of Docwra rock, the pleasure beach rides and 'sands of finest brown sugar'.[28]

The Parlour Outside

Fig. 33 Period postcard showing the interior of the Winter Gardens, Great Yarmouth, c.1903

Fig. 34 Period postcard showing the winter garden in the Royal Hotel,
Great Yarmouth, which opened in 1903

An amusement arcade and a roller-skating rink were added later. Also located on the east coast, and providing an easy escape for Londoners, Margate's winter gardens were opened in 1911. They comprised a large concert hall, four entrance halls, two side wings and an amphitheatre. Popular theatre and music, together with variety acts, constituted its mainly low-brow programme of events.

The concept of the palm court brought the domestic parlour and the public winter garden into a single space. Emulating aristocratic parlours and dining rooms, palm courts were formal spaces in which taking afternoon tea was often combined with listening to music and dancing. Palm courts permitted (mostly wealthy) people to feel at home and to interact with nature in spaces located within existing public buildings, from grand hotels to ocean liners to department stores. Topped by glazed skylights and filled with palms, they injected a sense of luxury and the atmosphere of a private parlour into semi-public settings. Though their inhabitants were away from home and keen to engage in social display in a public space, they also needed to feel that they were in familiar, unthreatening environments.

London's first hotel palm court appeared in the Langham Hotel, which opened in 1865.[29] The last years of the century saw several other elegant examples introduced into London hotels, among them the Savoy, which opened in 1889 and featured a tea lounge bordered by palms and other exotic plants.[30] The Carlton Hotel opened in 1899. In addition to its smoking, reading, dining, reception and retiring rooms, it also contained a vast palm court on its ground floor.[31] César Ritz, who oversaw the construction of the building, had requested Mewès and Davis, the firm that designed the hotel's interiors, 'to lower the level of the Palm Court [below that of the dining room] and introduce a flight of stairs for the effective entrances and exits of London's best-dressed ladies'.[32] In addition, 'a small balcony at the upper level gave a degree of privacy behind a screen of palms where the Prince of Wales could enjoy his regular visits from a vantage point over the Palm Court'.[33]

The furnishings and décor of the Carlton palm court were provided by Warings of Liverpool. Floral decorations were also introduced to reinforce the impression of luxury and public domesticity. In 1906, following the model successfully established in Paris in 1898, the London Ritz opened its doors, it too boasted an elegant palm court; in 1908 the Waldorf, a favourite among visiting Americans, opened, housing yet another impressive palm-filled room (fig. 35). As at the Carlton, the Waldorf's palm room featured a sunken area at its centre, which was accessed by a flight of steps.

Many of continental Europe's grand hotels, including Berlin's Central Hotel and Monte Carlo's Hotel Hermitage, also featured palm courts.[34] In the United States the fully fledged palm court was linked directly to the use of hotels by the newly wealthy as sites of conspicuous leisure and display. The palm garden attached to the Schlitz Hotel in Milwaukee was an early example. Built in 1895, five years after the hotel opened, it featured

PALM COURT, WALDORF HOTEL.

Fig. 35 Period postcard showing the Palm Court at the Waldorf Hotel, London (opened 1908), *c.* 1910

a barrel-vaulted ceiling and stained glass, and hosted an orchestra. San Francisco also had its share of late nineteenth-century luxury hotels, among them the Palace Hotel (fig. 36), which opened in 1875 and featured a plant-filled seven-storey atrium. The original hotel burnt down in 1906 in the fire that followed the earthquake of that year, but a new building was opened in 1909 and it too had an extravagant palm court at its centre. Most renowned of all the American examples, though, was the glamorous palm court in New York's Plaza Hotel, which opened in 1907. The Biltmore Hotel opened in the same city in 1913. Its tea room, or palm court, echoed the design of the main concourse of the Grand Central Railway Terminal, to which it was linked. Another of the many notable North American palm courts could be found in the lobby of Montreal's Ritz-Carlton Hotel, which opened in 1912.

Given that ocean liners were essentially luxury hotels on water it is not surprising that, by the early twentieth century, they too frequently contained elegant palm courts and provided extravagant tea dances for their (first-class) travellers. Examples could be found on the White Star Line's SS *Majestic* (launched in 1889), RMS *Olympic* (launched in 1910) and RMS *Titanic* (launched in 1912); the Hamburg America Line's SS *Amerika* (launched in 1905), SS *Imperator* (launched in 1912 and which later became Cunard's RMS *Berengaria*), SS *Bismarck* (launched in 1914) and SS *Johann Heinrich Burchard* (also launched in 1914, later sold to United American

Fig. 36 Period postcard showing palms in the dining room of the Palace Hotel, San Francisco (rebuilt 1909), c.1910

Lines and renamed *Reliance*); and the Holland America Line's SS *Nieuw Amsterdam* (launched in 1905).

Palm courts and palm-filled sitting areas were also introduced into department stores to encourage female shoppers to relax and to spend more time and money in them. London's Harrods featured a number of such spaces. As Erika Rappaport has explained, 'stores presented themselves as safe, pleasurable, and emancipating places for urban women'.[35] Adding plants introduced the security of the parlour. Selfridges, which opened in London in 1909, contained a luxurious Palm Court Restaurant that was destroyed by a bomb in 1941. In commercial settings the role of nature inside was to provide a relaxing, safe, but also stimulating and luxurious environment in which consumers could feel simultaneously at home and that they were 'out on the town', enjoying a break from chores and responsibilities. Plants succeeded in helping people cross the boundaries between reality and fantasy and between the private and the public spheres. As mass consumption expanded at the end of the nineteenth century, the role of indoor plants in encouraging that activity gradually took over from their earlier functions as religious symbols, educational tools and decorative objects.

Palms and other exotic plants filled many other semi-public spaces in the years around the turn of the century, where they helped to create 'home

Fig. 37 Period postcard showing the inhabitants of a convalescent home in the north of England, photographed in the winter garden, c.1920

from home' atmospheres. These included private clubs, residential schools, hospitals (see fig. 107), convalescent homes (fig. 37), mental asylums and sanatoria, in all of which the plants' task was to be therapeutic. The more content the patients, it was contended, the easier they were to manage, and both time and money were saved. Many photographers' studios also featured potted plants, in order to give the impression that sitters were at home, to relax them and to provide a photogenic setting (fig. 38).

Although many of the urban buildings described in this chapter – the leisure centres, department stores and hotels – housed modern-looking winter gardens of iron and glass, these structures were often invisible from the street. In keeping with the surrounding environment, the main language of these architectural structures was that of nineteenth-century historicism and eclecticism. Two routes out of that conservative approach were provided by the protagonists of the British Arts and Crafts movement and by European Art Nouveau architects and designers, who looked to the world of nature as a source of contemporary imagery. Unless conservatories were attached to their buildings, however, their efforts to represent flowers and birds, among other natural forms, in their designs for furniture and furnishings did not extend to the inclusion of living nature in their interior settings.

Fig. 38 Period postcard showing a couple surrounded by indoor plants, photographed in J. Busby Cox's Westgate Studio, Peterborough, c.1910

A little later however, several international architects and designers in search of a new language for the modern age took yet another route, embracing the metaphor of the machine and promoting functionality in a manner that, in some senses, recalled the belief in rationalism that had dominated eighteenth-century Enlightenment thought. In this new context, the affective concept of domesticity became increasingly problematic, and, because of their close associations with the concept of home, indoor plants and flowers were excluded from the dominant architectural and design discourses of the new era. However, as the following chapter will explain, they did not disappear entirely.

CHAPTER 4

Natural Modernism

The cactus sprouts where once flourished the aspidistra and the rubber plant.[1]
Osbert Lancaster

Although the ideas and forms of Victorian middle-class domesticity continued to spread through society in the years after 1914, both in and beyond the home, its rich symbolism was gradually diluted by the growth of secularism, by the increasing emphasis on (masculine) rationality, science and technology, and by progressive architects' negative association of the concept of home with bourgeois culture and the marketplace. By the inter-war years of the twentieth century domesticity had been relegated, for the most part, to the conservative spaces of the suburbs. As the urban masses came to dominate the public sphere in the aftermath of the First World War, the world of Victorian and Edwardian winter gardens, and the semi-public exoticism and luxury of the palm court, also disappeared from view.

At that time, a generation of forward-looking architects and designers working under the banner of what later came to be known as 'modernism' set out consciously to eliminate the concept of private domesticity from both their residential and their public buildings. Driven by a utopian desire to start from scratch, to modernise and democratise their buildings, and to eliminate clutter – which, in their minds, was synonymous with the stuffy world of their childhoods, upward aspiration, conspicuous display, excessive materialism and femininity – they sought to create rationally conceived, minimally furnished inside spaces that were extensions of their architectural frames and, ideally, indistinguishable from them. At its most extreme this approach led to the elimination of the interior in favour of a seamlessness between inside and outside spaces that denied the possibility of enclosure. The new aesthetic made a strong commitment to light and air, and drew

Fig.39 Hans Scharoun, Villa Schminke, Löbau, completed 1933. Photograph by Florian Monheim, 2012

heavily on the metaphor of the machine and the geometry and rationality these designers and architects believed to flow from it. The concept of 'home' came under scrutiny and was replaced by the term 'dwelling', a significant shift that prioritised functionality over the world of the emotions and the senses.

In spite of the general trend to eliminate all signs of home, the notion of a new domestic interior fit for the machine age was discussed by some writers. In 1929, for example, Dorothy Todd and Raymond Mortimer acknowledged that there was still a place for individualism in the age of mass production, but that, in interiors, its expression needed to eliminate unnecessary decoration and recognise the importance of light and openness.[2] Plants and flowers also continued to play a role in many modernist interiors. Although they existed in tension with the dominant rhetoric of progress and the metaphor of the machine, they provided a level of decoration that several modernist architects deemed appropriate. Arguably, nature inside contributed to a new model of modern domesticity, one in which plants and flowers played an increasingly important aesthetic role, reinforcing the modernists' preoccupation with the articulation of space, and softening the hard materials – concrete, steel and glass – that otherwise prevailed.

In 2007 Paul Overy became one of the first architectural historians to document the presence of nature in modernist interiors. He observed, for example, that 'In one of the early photographs of the entrance hall of the Villa Savoye, a vase of flowers can be seen placed on the plain hall table attached to one of the pilotis.'[3] This comment was unusual. Although flowers and plants were one of the few forms of ornament admitted into the modernist interior, they were studiously ignored by contemporary critics and by later historians of the movement, who have tended to focus on its more mechanistic elements.

Although the rhetoric that accompanied modernism ignored indoor plants and flowers, nature in its most general and abstract sense was addressed by several architects working in the inter-war years who embraced an organic version of modernism. Earlier in the century the American architect Louis Sullivan had described the creation of architectural form as emulating the process of a growing plant, which begins with a root, develops into a stem, and finally produces leaves and flowers. His pupil Frank Lloyd Wright followed in his path.[4] Natural laws with their own internal geometries provided, the advocates of the organic model believed, inspiration for the production and understanding of architecture. Later, the German architect and organic modernist Hugo Häring explained, 'If we prefer to search for shapes rather than to propose them, to discover forms rather than to construct them, we are in harmony with nature.'[5] Yet although, on one level, the approach of the organic modernists contrasted with the form of modernism that took the machine as a starting point, and for which strict geometry was a given, their work was still rooted in a form of abstraction.

Rather than seeing it as wild, dangerous and unruly, the modernists saw nature – especially as manifested in sunlight and fresh air – as benign, generous and helpful to human beings.[6] As a result, they often constructed their villas and apartment blocks in direct relationship with the more or less tamed gardens, grounds and landscapes that surrounded them and onto which they looked, taking great care to maximise the amount of sunlight entering them. The strikingly progressive, white, flat-roofed, ribbon-windowed buildings on the Weissenhof Estate, which was created as part of the Deutscher Werkbund's Stuttgart exhibition of 1927, were set on a hill that was covered with grass, trees and bushes. The architects involved considered that natural setting to be an integral part of their designs. Writing later about the Stuttgart project, Overy felt that '[Hans] Scharoun, [Adolf] Rading and [Richard] Docker integrated their houses with the landscape and terrain more effectively and unobtrusively than most of the other architects at Weissenhof.'[7]

The house that Hans Scharoun created for Fritz Schminke, the owner of a noodle factory, in Löbau, Germany, in 1930–33 was an exercise in organic modernism (fig. 39). Surrounded by extensive grasslands, formed into a garden by Herta Hammerbacher, the house was pushed into a sloping

Fig. 40 Eileen Gray and Jean Badovici, Villa E-1027, Roquebrune-Cap-Martin, completed 1929. Photograph by the author, 2011

site at an angle that gave its inhabitants a view across the garden and the fields beyond. A strong dynamic relationship was thereby created between the house and the garden.[8] The house also featured a winter garden on its ground floor. Ludwig Mies van der Rohe's Villa Tugendhat of 1929–30, built in Brno in the modern Czech Republic (the subject of the following chapter), also took full advantage of the garden that surrounded it, while Le Corbusier's Villa Savoye (1928), built in Poissy, France, hovered on pilotis over a relatively untamed piece of sloping land covered with grass and trees. A rectangular cut-out in the concrete wall of the villa's roof terrace acted as a frame, taking full advantage of the view of nature beyond. Le Corbusier's concept of the *promenade architecturale* also linked the landscape seamlessly to the villa, creating a dynamic dialogue between nature and culture. Le Corbusier had shown an early interest in the natural world: in 1910, for instance, he had drawn the orangery at Sanssouci, and a couple of years later he drew the example at Versailles. It was not until 1930, however, that he became interested in natural objects – including shells, bones, wood and flint – which he associated with a form of primitivism and which inspired a great deal of his later work.[9]

Eileen Gray's Villa E-1027 of 1926–9, built with Jean Badovici in Roquebrune, southern France, took full advantage of its situation close to the Mediterranean Sea, which provided the main view from the front of the house (fig. 40). When, in 1951, Le Corbusier constructed his little Cabanon next to Gray's house, he too was undoubtedly attracted by the site and its close links to nature. Across the Atlantic in Pennsylvania, Fallingwater, a house designed by Frank Lloyd Wright in 1935 for Edgar J. Kaufmann, took the idea of placing a building in a natural setting even further (see fig. 91). As Timothy Benton has observed, 'the design for Fallingwater brings out Wright's obsession with making a dwelling partake of the landscape in as intimate a way as possible. The house not only straddles a small waterfall but is open to the stream ... and bathed in the sounds of the moving water.'[10] No effort was made to hide the house in the nature that enveloped it; the goal was rather to create a visual dialogue between them.

The occupants of many modernist villas and apartments could engage directly with the natural world outside their buildings through the introduction of horizontal picture windows that framed picturesque slices of the exterior view. The 'pictures' thus created acted as substitutes for the works of art that might have otherwise adorned the walls. Writing of the United States in the years after the Second World War, the architectural historian Sandy Isenstadt has maintained that buildings were placed so firmly in their natural settings that those rectangular pieces of glass had come to be understood as a 'harbinger of the good life'. Most significantly in the context of taming nature and transforming it into culture, Isenstadt states that '"framing" was the learned activity that transformed *nature* into *landscape*'.[11]

Fig. 41 Walter Gropius, Masters' House, Dessau, completed 1926. Photograph by Alan John Ainsworth, 2018

Many modernist architects used a cluster of architectural features including balconies, verandas, terraces, roof gardens and external staircases, to extend their buildings out into nature, to let in light and sunshine, to facilitate seamlessness between inside and outside, and to avoid the sense of enclosure that had been the norm in most Victorian buildings. A house in Pasadena designed in 1908 by the local architectural partnership Greene & Greene exemplifies that seamlessness. Built as a winter home for David and Mary Gamble, three of the first-floor bedrooms of the Gamble House had exterior sleeping porches leading directly from them. Covered over, they were nonetheless open to the elements at the sides. Window boxes filled with trailing ivy were attached to the exterior walls as a means of linking the porches to the outside world of nature. Two decades later, Le Corbusier included the use of terraces and roof gardens in his 'five points of architecture', which he published at the time of the Deutscher Werkbund exhibition.[12] His Villa Stein of 1926 (on the outskirts of Paris) incorporated a complex system of them. Balconies were also widely used by many of the architects who created buildings for the Weissenhof Estate, Scharoun and Rading among them, as were stepped terraces, which made many of the buildings resemble ocean liners.[13] The emphasis was on outward-facing living, and on engaging with nature outside. One model for the approach was provided by contemporary tuberculosis sanatoria, where patients were exposed to the outdoor elements once a day.[14]

According to Overy, Scharoun's Villa Schminke was 'more balcony than house'.[15] It also featured an external staircase and numerous terraces that served to link the house to the garden. The design also took the association with ocean liners to a new level, including the addition of what looked like ships' railings. The houses built for the Bauhaus masters in Dessau were also exercises in indoor/outdoor ambiguity, so defined were they by their balconies, terraces and roof terraces (fig. 41). The same strategies were soon deployed all over Europe. In France, Robert Mallet-Stevens's Villa Noailles (1923–7) featured a sleeping terrace with electrically operated mosquito curtains, while in Austria, Richard Neutra's 1932 house for the Vienna Werkbundsiedlung boasted a roof terrace.[16] Many apartment blocks of the period also featured balconies, which acted as substitutes for gardens and in which potted plants and window boxes were common elements. Those introduced into the Siemensstadt residential estate in Berlin, masterminded by Scharoun and built between 1929 and 1934, were a case in point.

While siting a building in a natural setting, enabling inhabitants to view that setting from inside, and utilising various architectural features to blur the boundaries between the insides and outsides of buildings represented some of the ways in which the modernist interior was made to interact with the natural world, another strategy resulted in an even more intimate relationship between them. Several modernist structures were designed and built around trees that were already growing on the site. This particular architectural intervention took respect for the natural world one step further, demonstrating a reverence for nature in its growing state. The idea went back to the Crystal Palace, which, as we have seen, accommodated several large elms that were already established on its Hyde Park site. The act of enclosing trees represented continuity and disruption simultaneously. The trees could continue to grow, albeit within an artificial enclosure, and, although they were separated from other trees located outside the structure, they remained untouched and protected.

As the Crystal Palace had been a temporary enclosure, there was a logic to leaving the trees on site. The same could be said about the tree that was left on site when Le Corbusier designed his little Esprit Nouveau pavilion for the 1925 International Exhibition in Paris. By cutting a circle in the roof and allowing the tree to grow through it, Le Corbusier transformed the tree into an architectural feature. Since the pavilion was only temporary, that approach made some practical sense. However, between 1889 and 1895, Frank Lloyd Wright implemented a similar strategy in his own home in Oak Park, outside Chicago, in which he constructed a passage between the home and the studio so as to avoid disturbing a willow tree that was growing there. His gesture represented something somewhat different: a respect for nature in its wild state and a desire to blend it with architecture as sensitively as possible.[17] A similar impulse is apparent in Le Corbusier's later design for the Curutchet House in Argentina (1948–53), where the architect once again left a tree on site – an act that was, simultaneously, both non-interventionist and one of containment. In the Mill Owners' Association Building in Ahmedabad, India, designed in 1951, Le Corbusier went a step further still, leaving the façade largely open so that plants could grow on it, thereby letting nature itself act as the boundary between the interior and the exterior of the building.

Indoor winter gardens offered another means of bringing nature, in the form of plants and flowers, inside modernist houses. Although rooted in the attached Victorian conservatory, the strategy embraced by modernists brought nature even further into the interior. It recalled the approach adopted earlier in the century by Isabella Stewart Gardner, who had wanted her garden to be located at the very heart of her building. An organically shaped winter garden in the aforementioned Villa Schminke, filled with exotic plants, flowers and a water basin, was positioned next to the main living area, separated from it by a low curved rail.

Walter Gropius introduced winter gardens into two of the private houses he created in Jena: the Auerbach House of 1924 and the Zuckerkandl House of 1927–9.[18] On its ground floor the former featured a fully glazed, steel-constructed conservatory with three windows facing east, a south-facing double door, and three removable windows that interfaced with the dining room and could be concealed behind wooden blinds. Like Scharoun's Villa Schminke, the three-storey Zuckerkandl House was built into a steep slope. A glazed conservatory, topped by a terrace, protrudes from the façade of the middle floor. In an exterior photograph taken in 2009, exotic potted plants, including cacti and agaves, can be observed on its tiered shelves (fig. 42).

In 1927 the Viennese architect Josef Frank designed a conservatory in the house he built in Vienna for Anna Lang and her first husband, Robert.[19] He created yet another example for a residence at Wenzgasse 12, Vienna (1929–31), which he designed with Oskar Wlach. Although a modernist, Frank espoused a soft, textile-led, domestic aesthetic in his interiors, in which plants and flowers played an important part. The decorating company that he ran with Wlach from 1925 onwards – Haus & Garten – was responsible for many eclectic interiors that embraced nature in a number of ways. A late-1920s drawing for a living area that Frank created for Johann Bochynek, for example, featured a mixture of eighteenth-century and modern furniture styles as well as a small winter garden leading off the room through a door in a glazed wall. The architects Alfred and Emil Roth also positioned an indoor garden, containing a mother-in-law's tongue, in the ground-floor entrance of the apartments they built with Marcel Breuer for Sigfried Giedion in Doldertal, a suburb of Zurich (1935–6). They offset the garden with a large, abstract piece of organic sculpture created by the English sculptor Barbara Hepworth (fig. 43).

Winter gardens also found their way into many non-residential modernist interiors, especially ones that acted as homes from home. Recalling the spectacular palm courts of the early twentieth century, an extensive winter garden was introduced into the French ocean liner SS *Normandie*, which

Fig. 42 Walter Gropius, Zuckerkandl House, Jena, completed 1929.
Photograph by Nico Stengert, 2009

Fig. 43 Barbara Hepworth, photomontage showing her *Helicoids in Sphere* installed in the entrance hall of the Doldertal apartments, Zurich, designed by Alfred and Emil Roth with Marcel Breuer (completed 1936), 1938

Fig. 44 The winter garden on the ocean liner SS *Normandie* (launched 1932), 1935

was launched in 1932 (fig. 44). Filled with exotic plants, including palms, and tiered birdcages, it provided a glamorous, restful place in which travellers could sit in wicker chairs, take refreshment, and imagine they were in the tropics.

The European habit of using double panes of glass in windows in order to retain heat provided another perfect opportunity to bring plants inside, as the spaces between the window panes acted as miniature greenhouses. Two Italian modernist buildings included plants in this way. The first, the Electric House of 1930, was designed by Group 7 (Luigi Figini and Gino Pollini, in particular) for the Fourth International Triennial Exhibition of Modern Decorative and Industrial Arts in Monza (fig. 45). Corbusian in design, the gap between the two, 33-foot-long, L-shaped panes of glass at the front of the building – a shallow greenhouse, in effect – contained a row of cacti set into a long, thin bed of sand and stones. Their role was, once again, to create a sense of permeability between inside and outside. The reference to nature contrasted starkly with the central function of the building, which was to promote technologically advanced electrical appliances and showcase a range of new materials, including chromed metal, celluloid and linoleum.

The Villa Necchi Campiglio in Milan was designed between 1932 and 1935 by Piero Portaluppi for the sisters Gigina and Nedda Necchi and Angelo Campiglio, Gigina's husband (fig. 46). Although the style was more *novecento* than modernist, and the interior furnishings were overtly

Fig. 45 Interior of Group 7's Electric House, designed for the 4th International Triennial Exhibition of Modern Decorative and Industrial Arts, Monza, 1930. The principle designers were Luigi Figini and Gino Pollini. Photograph by Studio Bombelli

Fig. 46 The veranda room and window greenhouse in the Villa Necchi Campiglio, Milan, designed by Piero Portaluppi, completed 1935. Photograph by the author, 2011

Fig. 47 The living room of the Riihitie House (Aalto House), Helsinki, designed by Alvar and Aino Aalto, completed 1936. Photograph by Fred Runeberg, 1940s

luxurious – antiques, and artworks both old and new – the building embraced plants inside in a similar way to the Electric House. They were placed in the miniature greenhouse created by the space between the panes of the large double-glazed windows in the veranda room, which was comfortably furnished like a living room with upholstered chairs and a coffee table. The captured plants foregrounded the view of the garden that lay beyond them. Looking out onto the garden involved taking in two layers of greenery, the first comprising the small cultivated plants in the space between the window panes, and the second consisting of the larger-scale, wilder nature beyond.[20]

In order to avoid loose clutter in their interiors, certain modernist architects liked to build furniture into the structure of their buildings. One of the ways in which they included plants in their indoor settings was by creating built-in planters and shelves to demonstrate that, rather than being mere afterthoughts, the plants were integral components of their interior schemes. Frank Lloyd Wright used built-in planters on the exterior of many of his buildings. The Robie House in Chicago (1910), for example, had 'planters hidden in nearly every horizontal ledge'.[21] A decade later his Hollyhock House (1919–21), built in Los Angeles in the Mayan Revival style for the oil heiress Louise Aline Barnsdall, who had requested a 'half house, half garden', featured numerous rectangular concrete planters situated both inside and outside the building (see fig. 64).[22]

Although the Finnish architect Alvar Aalto embraced Corbusian principles in many of his buildings, working alongside his wife, Aino, he also developed an aesthetic of domesticity that embraced soft elements, among them natural materials, textiles and indoor plants. The Riihitie House in the Munkkiniemi area, just outside Helsinki, which the couple designed and inhabited together, has been described by Juhani Pallasmaa as 'a convincing compromise between the radical tendencies of invention and novelty, and the romantic concern for tradition and familiarity' (fig. 47).[23] Completed in 1936, it contained built-in wooden planters positioned at the bases of the windows in the living and dining rooms, which were filled with potted plants. Ivy was trained to cover the fireplace wall. The house also featured a roof terrace, a balcony and an exterior patio that linked the house to the garden. Aino Aalto, who made many of the textiles for the house, was fond of plants, especially cacti and agaves, and it is very likely that she was responsible for their inclusion.[24] In 1937 she designed and had made two organically shaped ceramic plant pots for the house, which were used on the terrace.

The rooms that the German architect and designer Marcel Breuer created in 1927 for the Berlin apartment of the theatre producer Erwin Piscator, are often presented as classic examples of modernist interior design (fig. 48). Sparsely decorated and furnished with metal and glass items, they were strikingly minimal. And yet, in the living area, positioned above a radiator on the wall that separated that space from the dining room, a small wooden shelf hosted a row of tiny potted cacti, a plant that Christopher Wilk has claimed is often considered '*the* plant of the

Fig. 48 The living room in the Erwin Piscator Apartment, Berlin, designed by Marcel Breuer, 1927. Photograph by Sasha Stone, 1928

Fig. 49 'Living Room with Cacti', an illustration from H. Hartl, *Modern Interiors in Colour* (Stuttgart: Julius Hoffmann, 1929), plate 84

modernist interior'.²⁵ The choice of that exotic plant, made possible by the advent of central heating, was undoubtedly driven by the fact that its small, simple, sculptural forms suited the look and feel of modernist interiors. The cactus also had a wide set of cultural references. Given its links to Mexico and the desert it suggested a place unlike northern Europe, one that was sun-drenched and associated with primitive art forms. Often used evocatively by novelists of the era, D. H. Lawrence among them, and painted by such artists as Charles Sheeler, in the age of mechanisation and mass production the cactus represented the exotic 'other', replacing the imperial palm of the previous century (fig. 49).²⁶

Free-standing potted plants were also added as decorative features in many modernist interiors, both private and public. Single potted plants were undoubtedly sometimes included as afterthoughts – especially when a promotional photograph was needed – but, as in Victorian interiors, they frequently constituted core aesthetic elements in decorative schemes. In the context of modernist architecture, however, there was much less clutter, and greenery was carefully displayed to offset the otherwise stark, minimal spaces it inhabited. As fewer plants were being used, their positioning was all the more powerful and strategic.

Books on the new interior illustrated inside spaces from all over the world, both private and public, many of which contained plants and flowers. Herbert Hoffmann's *Modern Interiors in Europe and America* of 1930, for example, included, among many others, an image of a Parisian hotel's entrance hall designed by Pierre Chareau, in which a potted dracaena was positioned on a shelf above a radiator; one of a bedroom in a villa on the French Riviera designed by Georges Djo-Bourgeois and his wife, Elise Bourgeois, which featured a vase of daisies on a chest of drawers; and an image of a hallway in a Dortmund building designed by Emil Pohle, which featured a large rubber plant on a side table.²⁷

The stands and pots that contained indoor plants also exercised the minds of designers in the inter-war years. For instance, the craftsman and metalworker Louis Dalbet collaborated with the architect Bernard Bijvoet to create striking metal containers of varying heights for Pierre Chareau's Maison de Verre (1932), which was designed for the fashionable Parisian gynaecologist Jean Dalsace.²⁸ Grouped together into clusters, the plants in these containers provided a contrast to the geometrical regularity of the house. One group of potted plants, which included a large dracaena and a weeping fig, was positioned just inside the large glazed window at the front of the house.

As was the case with growing plants, arrangements of cut flowers in modernist interiors were generally restrained and inconspicuous. That was largely a result of the strong influence of Japan, mentioned in Chapter 2. Frank Lloyd Wright had been among the first to advocate arranging flowers in the Japanese (*ikebana*) style. In his home in Oak Park, for example, he included what he described as a 'weed holder', which he had designed himself

Fig. 50 The lounge area of the German airship LZ 127 *Graf Zeppelin*, designed by Bernhard Pankok, completed 1928

and in which he had placed a single birch branch. It was a form of decoration that he repeated in several of his buildings, including the Martin House in Buffalo (1904). While bunches of flowers in vases very often denoted bourgeois domesticity, because the pared-down arrangements of modernism were less flamboyant than their more ostentatious Victorian antecedents, they were apparently less likely to be associated with social aspiration.

Whether suggested by designers or introduced by inhabitants, flower arrangements, used sparingly, became a regular feature in modernist interiors. They appeared in several of the houses on the Weissenhof Estate, including those designed by Rading, Frank and Gropius. As we have seen, they were usually the only decorative item admitted, their link to nature distancing them from the numerous non-natural bibelots used in Victorian interiors. A photograph of Marianne Harnischmacher in the living room of her 1932 house in Wiesbaden, designed by Marcel Breuer, depicts her sitting reading next to a table on which stands a vase filled with an arrangement of twigs and branches. Even the interior of the German airship *Graf Zeppelin*, designed by Bernhard Pankok in 1928, included a vase of roses placed on a side table in an attempt to make the space look homely and to relax its fearful passengers (fig. 50).

Through the inter-war years in both Europe and the United States, the fashion for flower arranging in the Japanese style was disseminated to an audience of amateur female home-makers through home-making advice books. In 1929 Anne Lamplugh illustrated arrangements of horse chestnut and other blossoming branches.[29] 'Just three sprays of blossom make a very lovely vase if arranged in a modification of the Japanese manner', she explained. She continued: 'The most beautiful results are produced with the slightest materials', while suggesting that 'an even more modified application of this Japanese effect can be obtained by using one branch of blossom wedged among stones in a large dish, and stood upon a low stool in a corner, or in the angle formed by some piece of furniture and the wall'.[30] Echoing Frank Lloyd Wright, the fashionable and highly authoritative English flower arranger Constance Spry offered similar advice in her book *How To Do the Flowers* (1938), explaining that 'decorative effect may be achieved with simple materials, with a few leaves and grasses or weeds'.[31] The anonymous author of *The Home of Today* of the same year, an advice book published by the *Daily Express* newspaper, also declared that 'It would be a good thing if we would take lessons from the Japanese and not crowd our flower vases or mix our colours.'[32] In 1947 the writers of an American publication on the same theme claimed that 'the type of flower arrangement which may justly be defined as "modern" is a hybrid resulting directly from the East and the unfettered, robust, colourful arrangements of the West', implying that by the post-war years modernism had met decoration half way, and the result was a compromise between Eastern and Western aesthetic principles.[33]

Although the indoor planting and flower-arranging styles of the Victorians and modernists differed dramatically, the Victorians linking nature directly to God while the modernists exploited its aesthetic properties, both groups shared a belief in the power of nature to bring additional life into spaces that were otherwise conceived, made and inhabited by human beings alone. In modernist buildings, nature was used strategically to erode the distinctions between inside and outside and thereby liberate interiors from all trace of enclosed domesticity. At the same time, by softening indoor spaces, providing a form of decoration and undermining the mechanistic model of modernism, some modernists were also reclaiming the domestic interior. That seeming tension was visible in several iconic modernist buildings, not least the subject of the next chapter, the house that Ludwig Mies van der Rohe built for the Tugendhat family in Brno.

CHAPTER 5

The Garden Inside: The Villa Tugendhat, Brno

Every plant and animal demonstrates a different way of life ...
What we share with them is at least as profound as our differences.[1]
Joseph W. Meeker

As explained in the previous chapter, in order to escape the enclosed, inward-looking, symbol-laden interior that had defined Victorian domesticity, the designers of the inside spaces of inter-war modernist dwellings pursued a set of innovative formal and spatial strategies. Though there were tensions, plants and flowers often played a key role in that process. Whether they were included, consciously or subconsciously, to demonstrate a level of continuity with nineteenth-century domesticity, or whether they were part of the desire to create a new aesthetic language for the modern dwelling, is up for debate. Most accounts of inter-war architectural modernism have ignored their presence in favour of narratives that have foregrounded aesthetic formalism, technological innovation and socio-political ambitions.[2] Some modernist buildings were openly declared by their inhabitants to be difficult to live in – Ludwig Mies van der Rohe's Farnsworth House (1945–51) in Illinois, and Le Corbusier's Villa Savoye (1928–31) in Poissy, France, come to mind – while others were reportedly easier to inhabit. Depending on the circumstances, the presence of plants either added to the difficulty or alleviated it.

As well as developing new housing schemes for the working classes, many modernist architects also created residences for wealthy middle-class clients who sought a modern lifestyle free from possessions and clutter. Those progressive clients aspired to a liberated existence that focused on an outward orientation and a sense of escape from the claustrophobia and enclosure associated with the past. Several of these dwellings contained natural elements. One such was Mies van der Rohe's Villa Tugendhat in

Fig. 51 Rear view of the Villa Tugendhat, Brno, designed by Ludwig Mies van der Rohe, completed 1930; restored and photographed 2012

Brno in the modern Czech Republic, built between 1929 and 1930 (and extensively restored between 2010 and 2012), as the family home of Fritz and Greta Tugendhat.

Mies's villa has been seen by many architectural historians as a supremely successful example of a residential structure that allowed space to flow uninterrupted; that encouraged an idea of modern family life that did not, at least on one of its floors, require a set of self-contained rooms; and that achieved these goals using advanced technologies and modern materials.[3] Much attention has also been paid to the architect's inclusion of rich, highly textured materials – onyx, travertine and a variety of exotic woods – which have been understood as markers of the modern luxury deemed appropriate in that middle-class setting.[4] They can also be understood as representations of the natural world in a building largely constructed from the non-natural, hard materials of modern industry – steel, plate glass and concrete among them.

The fact that Mies chose the site carefully has received much comment, and the sloping garden that lies behind the house has been described as a dramatic setting for the striking, modern statement at its summit (fig. 51).[5] Less has been said, however, about the deliberately engineered

Fig. 52 Flower beds at the rear of the restored Villa Tugendhat. Photograph by the author, 2014

Fig. 53 Exterior view of the winter garden on the east side of the restored Villa Tugendhat. Photograph by the author, 2014

continuities between nature and the built structure, about the fact that tamed nature, in the form of the cultivated flower beds immediately outside the building, gave way to a wilder version of nature in the form of trees and bushes as one moved away from the villa to the periphery of the garden (fig. 52). That gradual shift from tamed to wild suggests a carefully controlled transition from culture to nature that reinforced the close relationship between them.

Descriptions of the terraces at the Villa Tugendhat have emphasised their role in ensuring that space flows freely from the inside to the outside of the house, and back again, thereby creating an inside/outside ambiguity rather than an enclosed, domestic interior. Much less has been said about how the terraces also provide opportunities for nature – from the climbing greenery covering the pergola that shielded the children's play area on the upper terrace, to that forming a screen behind the curved bench also positioned on that level – to come close to the building. Largely ignored, also, has been the sizable winter garden filled with exotic plants that flanks the entire east side of the building and brings nature right inside it, complementing the exoticism and rich textures of the other natural materials used in the design (fig. 53). A modernist version of the attached Victorian conservatory, it is linked to the house all along one of its sides, which makes it feel particularly closely integrated into the building and renders it more visible from inside. Similarly unnoticed have been the numerous potted plants positioned at strategic interior locations, which are visible in photographs taken throughout the villa's life, and the vases of flowers, also clear in many archive photographs, which were presumably positioned by Greta Tugendhat (or one of the servants). As well as providing an immediate exterior setting for the Villa Tugendhat, nature, in the form of plants and flowers, made a significant contribution to its inside spaces.

The functions and meanings of the nature in and around the Villa Tugendhat are more difficult to uncover, however. On one level nature was clearly used to reinforce the novel spatial strategies that enabled this remarkable architectural structure to repudiate the past. On another, perhaps, it served to consolidate the villa's relationship with the past and to introduce into it a degree of conventional middle-class domesticity.

The Villa Tugendhat has been extensively documented and analysed by architectural historians, partly because, in spite of its rich history, it still exists relatively intact.[6] Also, countless archive photographs of the villa exist, from those taken by Rudolf de Sandalo in 1930, the year of its completion, to others taken following its restoration in 2012. For the most part, those photographs are unpeopled studies in architectural form. The only exception is a set of photographs taken by Fritz Tugendhat between 1930 and 1938, which depict members of the family in the villa's spaces and communicate a sense of family life as it was lived in them.[7] The more formal photographic studies of the villa reinforce what have usually been seen as its most innovative features: its dramatic use of the open plan (in the middle floor); the strong sense of spatial continuity between inside and outside; and the importance of spatial flexibility manifested in the inclusion of movable screens and fabrics that could be used to create a variety of intimate private spaces as required.

The concepts of fluidity of space and transparency have been fundamental to understanding the Villa Tugendhat. They contribute to the idea that the architect wanted to design an interior that could not be defined merely as a series of rooms (at least on one floor). This strategy was facilitated by his decision to construct the building around a series of steel columns supporting a steel frame, rather than on weight-bearing walls.[8] The columns liberated the interior and exterior walls from their supporting function, allowing the inclusion of expanses of plate glass which, in turn, let in large amounts of light. A structural, engineering focus has dominated most analyses of the building, and many of its innovative technological features have been discussed at length, from the huge electrically controlled opening window in the living area to the use of exchanged air for heating and cooling its inside spaces. There have also been accounts of the furniture,

Fig. 54 Ludwig Mies van der Rohe and Lilly Reich on board an excursion boat on the Wannsee, near Berlin, 1933. Photograph by Howard Dearstyne

much of which was designed with open steel frames so as not to block the eye as it took in the villa's sophisticated and complex interior spaces.

An architecturally, technically, structurally and materially oriented account of the spatial configuration of the Villa Tugendhat is not the only way in which the inside spaces of this complex dwelling can be read and understood. Nor, indeed, was Mies van der Rohe its only creator. Its softer, more portable and ephemeral features – the textured materials, the textiles, the views outside, the play of light and shadow on its surfaces, the plants – also contributed significantly to its spatial and psychological novelty and sophistication. They were the results of various creative inputs: those of the architect; those of the designer of its textiles and colour scheme, Lilly Reich (fig. 54);[9] those of its garden designer, Markéta Roderová-Müllerová; and those of its inhabitants between 1930 and 1938, the members of the Tugendhat family. Given their relative instability and, sometimes, immateriality, the roles of the softer, transient aspects of the interior are inevitably harder to pin down. Arguably, the hard and the soft strategies worked together to reinforce the radical spatial ambitions of the architect and to provide a level of ambiguity regarding the villa's domestic character.

The Garden Inside

Located at Schwarzfeldgasse 45, the Villa Tugendhat sat, and still sits, on a suburban road on the outskirts of Brno, adjacent to several large, opulent, nineteenth-century, middle-class villas. Only one of its three levels can be seen from the street. The façade is covered in white stucco and features a sequence of opaque glazed windows, surrounded by metal frames. It curves round into a semi-circle at its west end. Today, three plants in square planters sit outside the glazed drum that conceals the entrance, softening the industrial severity of the standardised metal-framed panels of opal glass – materials that Mies transferred into his residential dwelling from the commercial arenas of store displays and exhibition spaces. A double-door garage is also visible at street level. The most striking feature of the villa, seen from that angle, is the void between the entrance drum and the garage, through which a framed view of Brno can be glimpsed. An exterior terrace runs from the front to the rear of the villa on that level, helping to emphasise its strong horizontality.

In 1930, on walking from the terrace through the concealed entrance behind the drum, the visitor entered a vestibule that featured a tiled floor of Italian travertine, strikingly marked with fibrous patterning, and a ceiling-to-floor dark palisander door positioned alongside a set of four wooden panels. A row of potted mother-in-law's tongues (modernist favourites) sat on a small travertine shelf fixed to the wall to the right of the vestibule. The small, regularly positioned plants with their sharply defined forms led the visitor's eye to the entrance door, in front of which two tubular-steel-framed cantilevered chairs were combined with a small side table to form a waiting area for guests.

To the left, at the top of the staircase leading to the floor below, a complex set of planes, surfaces, forms, materials and spaces converged. To add to that complexity, one of the internal, chromed-steel-covered columns passed through the midst of that convergence. The meeting point of the horizontal floor, which stopped sharply at a right angle as the stairwell fell away beneath it, and the vertical column was made even more complex by its proximity to the end point of the glazed drum and the presence of a rail, made of two horizontal rows of steel bars, that acted as protection from the otherwise open stairwell. An early photograph depicts a potted maple strategically placed on the floor at the point where all those elements met, softening the hard geometry and surrounding materials but also, arguably, providing a visual resolution to, or perhaps a distraction from, the multiple materials and the spatial complexity of the verticals, horizontals, curves, straight lines, masses and voids that came together at that point (fig. 55). The maple in question was fairly large. Its soft full leaves and loosely defined form provided a counterpoint to the formal complexity of what had undoubtedly been an architectural challenge. What may have seemed a minor addition (even an afterthought) already suggests how nature was being used in the villa as a means of solving complex visual problems and of enhancing the

Fig. 55 A potted maple positioned at the head of the stairs on the first floor of Ludwig Mies van der Rohe's Villa Tugendhat, 1929-30. Photographer unknown

Fig. 56 A potted maple in the restored Villa Tugendhat helps to recreate the building's original layout. Photograph by the author, 2014

building's architectural formalism. In recognition of its architectural and compositional importance, a similar plant was positioned in the same place in the restored villa in 2012 (fig. 56).

One wonders who put the plant there in 1930. Did the architect see its presence as important? Did the photographer, de Sandalo, feel it was a photogenic requirement? Or did a member of the Tugendhat family feel that it provided a picturesque punctuation mark that was needed in that specific location? On one level, given that interiors are always put together by multiple agents, those questions are irrelevant. What is important, rather, is to recognise the maple's primarily formal function in that architectural and interior setting. Not only did it reinforce the spatial strategies at play, it also added texture, colour and decoration to the interior. It may also have brought a level of conventional domesticity into that otherwise technologically progressive space.

Although it would have been structurally possible, the top floor of the villa was not designed on an open plan. It consisted, rather, of a set of private spaces – bedrooms for the Tugendhat family and the children's nursemaid. There was little that was radical about those spaces. Instead they were presented as conventional private rooms such as could have

The Garden Inside

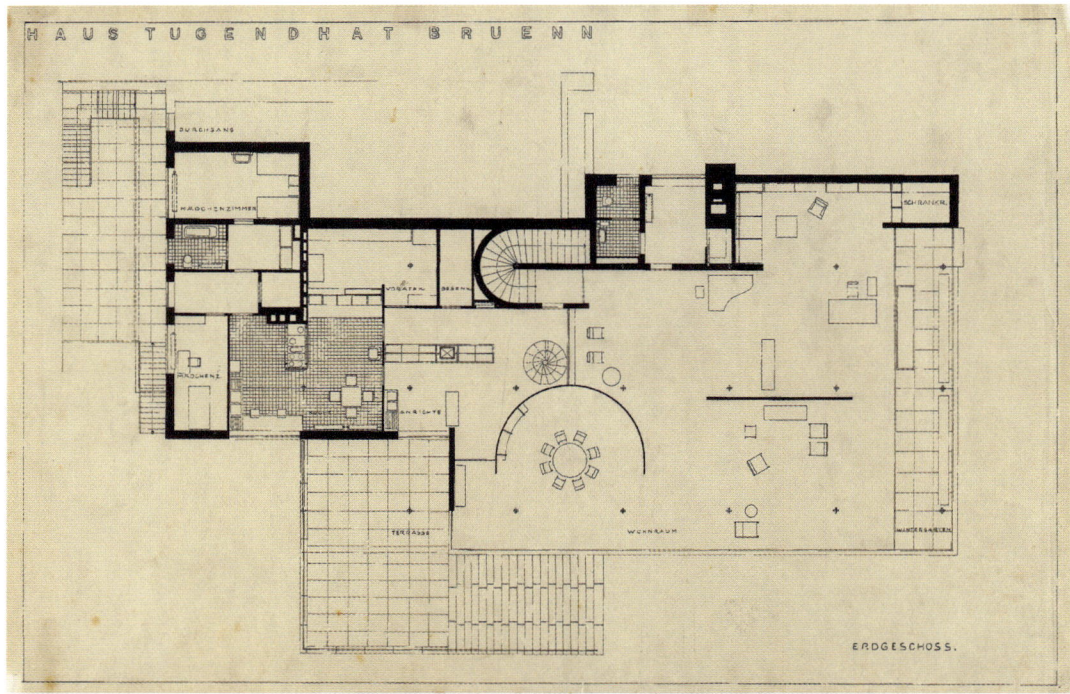

Fig. 57 Ground floor plan of Ludwig Mies van der Rohe's Villa Tugendhat, 1929–30. Ink and pencil on tracing paper, 24.5 × 38.5 in.

been found in any dwelling of the era. The rooms were particularly light and bright, however, and contained some very novel items of furniture, again made of exotic woods. They were designed by Mies, but undoubtedly with some assistance from Lilly Reich, who worked in his office at that time on his designs for furniture.[10] The windowsill of Fritz Tugendhat's room hosted a row of three small succulents of different types. They provided a touch of exoticism, as well as emphasising and reinforcing the role played by the large plate-glass window that framed the view of the garden outside. That frame helped both to dramatise the exterior view and to create a sense of inside–outside continuity, reinforcing the fact that Mies designed the villa from the inside out, rather than from the façade in.

The room led on to a terrace where more plants were in evidence, acting as transitions to those in the garden beyond. The indoor examples helped to initiate the movement from tamed to wild nature that has already been mentioned as a feature of the garden, and to link a private room intended for sleep and study to the world outside. In 2012, three small pointed succulents of the same species were positioned on the windowsill, suggesting, once again, that the role of plants in that location was still regarded as significant. Greta Tugendhat's room, which was next door, also led onto the terrace, and it too had windows that framed the

Nature Inside

Fig. 58 View of the living room in the Villa Tugendhat, with the winter garden in the background, 1929. Photograph by Atelier de Sandalo

striking natural scene outside. A small vase of flowers adorns a tubular-steel side table in an early photograph of the room. In 2012 a larger equivalent was added, undoubtedly to soften the otherwise hard materials in a space intended for habitation by a woman.

Descending the staircase to the floor below, the full impact of open planning became immediately apparent as an open space measuring 80 by 55 feet came into view (fig. 57). With the exception of the kitchen and the staff's living quarters, situated at the northwest end of the villa, the middle floor consisted of a single open living space. It was an area for family conversations, dining, entertainment, leisure and social interaction. Specific areas were partly defined by flexible textile screens.[11] The furniture was also positioned to support particular activities. The main furniture groupings included a round reception table with four specially designed tubular-steel-framed chairs at the entrance to the living space in front of a light wall; three Barcelona chairs (as they have come to be known) and a matching ottoman (all 1929), and three so-called Tugendhat chairs arranged around a Tugendhat coffee table (all 1930) and positioned next to a Moroccan onyx screen in the main living area (fig. 58); a dining area with an expandable rosewood table and Brno chairs (1930), enveloped by an enormous semi-circular Makassar ebony screen that echoed the form of the bench outside on the terrace;

The Garden Inside

Fig. 59 View of the study and living room in the Villa Tugendhat, with the winter garden in the background, 1929–30. Photographer unknown

a library area with a pigskin-covered sofa (1930) and three tubular-steel-framed Brno chairs (1930) arranged around a bridge table; and a rear living area with a large desk and two cantilevered chairs with armrests (fig. 59). A single metal-framed chaise longue (1931), upholstered in red fabric, was also included. Accompanied by a small side table, it offered an opportunity for private contemplation or quiet reading.

Those furniture groupings were placed on a neutral white linoleum floor and, with the assistance of several rugs, defined the main settings. The permeability of the border between the interior space and the outside world was reinforced on the south side by full-length windows, giving an impression that the interior extended beyond its actual physical boundary. The metal frames of the huge plate-glass windows controlled that view as if it were a set of large pictures on the wall. To render the effect of transparency even more dramatic, the two middle windows could be made to slide down into a groove that reached into the floor below, thereby creating real spatial continuity between the inside and the outside. Sitting inside, the inhabitants had a sense of being outside. Transparency became translucency when the vast see-through curtains were pulled across the picture windows. Opaque curtains on rails were also used to hide the winter garden from view, while another could be pulled across in front of the piano to turn the library area into a more private, contained space.[12]

While the uninterrupted views gave the impression of nature penetrating the interior, the winter garden that flanked the entire eastern end of the building literally brought nature inside. Full of exotic potted plants (including a striking cycad visible in archive photographs), luxuriant ferns and several other examples of greenery, which inevitably changed over the years, that huge glass container presented the unruliness of a jungle to those sitting on the chairs facing it. Juxtaposing the calm of the understated and sparsely decorated living space, the visual chaos of those thrusting evergreens was hard for the eye to take in. However, they were contained within glass walls and therefore prevented from invading the serene living area. Yet they were not so much tamed as controlled. Like ferns in a Wardian case in a middle-class nineteenth-century parlour, their link to the wilderness was enjoyed but not permitted to disturb the domestic space they inhabited. One could, however, enter the villa's winter garden and walk among the plants on a little path constructed for that purpose (figs 60 and 61), just as one could go into an attached domestic conservatory of the nineteenth century. This also allowed the plants to be tended, aided by the presence of water in a small pool served by a tap linked to the water supply. The pool added another natural element to the setting, reinforcing nature's presence in the building.

The villa's winter garden served many functions. On the simplest level it hid the interior from the street outside. On another it provided an intermediate, or transitional, space between the tamed interior and wild nature outside, reminding us, once again, of the architect's decision

Fig. 60 Perspective view of the winter garden in Ludwig Mies van der Rohe's Villa Tugendhat, 1929-30. Pencil on paper, 10.9 × 21.8 in. Kunstbibliotek, Staatliches Museen zu Berlin

Fig. 61 Inside the winter garden in the restored Villa Tugendhat, looking out to the willow and lawn beyond. Photograph by the author, 2014

to work from the inside outwards. Like the flower garden outside that bordered the façade of the villa, it offered a staging post between the tamed and the wild. It was part of the inside and part of the outside. Full of exotics, in winter it provided a layer of green against the white of the frequently snow-covered garden outside. Like the conservatory attached to the nineteenth-century middle-class home, it brought nature (almost) indoors; indeed, it did so more effectively than the conservatory, because the boundary-free nature of Mies's interior rendered it clearly visible from many positions in the living space, adding strong colour, pattern and texture to that setting. The visual complexity and controlled disorderliness of the plants in the winter garden, which were transformed by light falling on them, were complemented by the rich and seemingly random grain on the onyx wall. In both cases, nature offered Mies an opportunity to introduce decoration and texture into an otherwise minimally ornamented interior space.

Above all, the plants in the winter garden presented a riotous wall of green (or, rather, a range of greens) in that subtly toned interior. The green was picked up by the leather upholstery on the Barcelona chairs, one row of which sat with their backs to the vegetation. According to Greta's daughter, Daniela, 'Lilly Reich was largely responsible for the choice of colours and fabrics.'[13] Dominated by shades of white, off-white, black and silver chrome, the palette featured strong colour highlights in the form of the green plants and leather upholstery and the ruby red of the single chaise longue. Greta Tugendhat's flowers also offered temporary dashes of vibrant colour. Along with the multiple woodgrains, the travertine and the onyx, the leaves of the plants also provided pattern as well as texture, which was picked up again by the textile surfaces of the rugs and curtains. With the exception of the oriental rug in the library, which came from the Tugendhats' previous home and which added an obvious level of domesticity, the addition of those patterns and textures was probably Reich's work. The stone in the garden immediately outside the villa contributed to the effect, providing a rougher, untamed texture to join those inside.

The containment of uncontrolled nature achieved by putting plants within a large glass box added an important element to the villa and is one of its most interesting, and perhaps most surprising, features. It is also one of its least documented aspects. Spatially, the winter garden reinforced the sense of ambiguity, permeability and translucency that was so much a part of this novel project, adding another layer to the vista and providing an important bridge to outside. The presence of tamed and semi-tamed nature inside the villa directed the eye to look outwards at nature growing wilder by increments in the near, middle and far distances.

Archive images of the Villa Tugendhat show vases of cut flowers placed on a range of surfaces (a practice that continues in the restored building). Although these are the most ephemeral of all the interior components

Fig. 62 Indoor plants and flowers feature in Ludwig Mies van der Rohe and Lilly Reich's Exhibition House at 'Die Wohnung unserer Zeit', German Building Exhibition, Berlin, 1930–31. Photographer unknown

under discussion, they represent another way in which nature was introduced into the villa's inside spaces. They also added a level of domesticity. Greta's daughter has confirmed that her mother brought flowers in from the garden on a regular basis, and that her choices were notable for their modesty. She usually selected small posies rather than the exuberant bunches of flowers one would typically have found in a Victorian parlour.[14]

Small bowls of roses appear in a number of photographs: one on the tubular-steel-framed table in Greta's bedroom; another on a wooden table in the nursemaid's room; another on a little tubular-steel-framed table to the rear of the dining area; another on the dining table; another on the credenza beside the six Brno chairs; yet another on the sideboard behind the onyx wall; and a larger one on the piano. While they may have been placed there to enhance the picturesque or photogenic nature of the interior, Greta probably included them to introduce an ambiance of domesticity into the otherwise unrelentingly modern space.

The exotic plants in the Villa Tugendhat were not the first to have been brought inside a space designed by Mies. In 1927 he had collaborated with Lilly Reich on an interior for the Werkbund exhibition in Stuttgart.

Fig. 63 Potted plants in the lobby of the apartment building at 860–880 Lake Shore Drive, Chicago, designed by Ludwig Mies van der Rohe, completed 1951. Photographed before 1969

Matilda McQuaid has confirmed that 'the official catalogue listed both Reich and Mies as designers'.[15] The space became known as the Glass Room because of the extensive use of mirroring on its walls. One of its most interesting features, in the context of this study, was a row of three large potted plants – among them a large rubber plant and a dramatic agave – located on one side of the room, a precursor, without doubt, of the Villa Tugendhat's winter garden.[16] In 1930–31 Mies and Reich worked together again on an exhibition project in Berlin, 'Die Wohnung unserer Zeit' (The Dwelling of Our Time), which featured an ideal living space once again enhanced with potted plants and flowers (fig. 62). Nor was the Villa Tugendhat the last of Mies's buildings to contain plants. The lobby of his later Lake Shore Drive apartments in Chicago (1949–51) also featured several examples (fig. 63).

While Greta Tugendhat was clear that Reich selected the textiles and the colour scheme for her home, she did not attribute the choice of a conservatory to her. Instead, she explained that her husband was responsible for creating 'a genuine greenhouse in the conservatory'.[17] However, although Fritz Tugendhat clearly planted and nurtured its contents, the initial inclusion of the conservatory was not necessarily

The Garden Inside

his idea. Lilly Reich may also have been a key contributor, for the plants can be seen as having a softening and decorative role similar to that of the other aspects for which she was responsible, as well as adding to the exotic effect of the richly textured woods and other natural materials that she almost certainly helped to choose. The green provided by the plants could well have been her idea, too, at least in part.

The use of exotic plants and luxury materials brings one back to the question of conventional domesticity in the context of the Villa Tugendhat. Both Fritz and Greta Tugendhat came from families of wealthy industrialists. Rather than opting for a display of traditional opulence, by employing Mies to design their home they chose a more radical path. However, it appears that the villa did perform the role of home for them during the eight years in which they inhabited it. Greta, for example, claimed that the open plan allowed her both privacy and the possibility of being part of a totality.[18] Her decision to include her piano and her oriental rug, as well as her assiduousness in ensuring the continued presence of fresh flowers, suggest that she felt the need to add elements of conventional domesticity to the villa, and found ways of doing so. In addition, even though the pieces themselves were made from tubular-steel frames rather than horse-hair upholstery, the furniture arrangements facilitated family activities and social interaction. Nothing seems to have prevented the Tugendhat parents and children from enjoying family life in their villa, and, indeed, several of its features seem to have supported a traditional domestic existence.

In the debate about the villa's habitability, or the extent to which it permitted domesticity – a debate initiated as early as November 1931 in the Werkbund's journal, *Die Form* – the role of the plants and flowers is central.[19] They undoubtedly brought to the interior an exoticism and a sense of bourgeois luxury, both of which had been indicators of nineteenth-century domesticity. In addition to the possible addition of a conservatory, in a nineteenth-century home potted plants would have been distributed freely throughout the living spaces, adding to the sense of clutter, while also, paradoxically, providing a means of knitting everything together visually. In the Villa Tugendhat, in addition to the specimens located near the staircase, on a handful of shelves and on bedroom windowsills, the most dramatic plants were contained behind a large sheet of plate glass. That created a view, observable from many locations in the open living area, that functioned like an enormous painting or cinema screen, bringing tamed nature into the interior, but, at the same time, restraining it and not letting it invade. The glass prevented the family from engaging with the winter garden with any sense other than sight (unless they chose to go inside it), while its dramatic impact added a level of visual theatre to the interior.

That restraining of nature prevented it from becoming part of the interior itself. However, the fact that family members could open a door and walk directly into the winter garden meant that they could engage directly with it whenever they chose to. There was no longer a sense, as there had been in Victorian homes, that God was present there (though the villa has been described as having a spiritual quality), that the winter garden formed an important element within the children's education, or that it functioned as a marker of social status, yet the proximity to plants, the little path and the small pool of water clearly offered a moment of escape, of contemplation and of calm that undoubtedly enhanced the well-being of visitors to the conservatory and of those seated in the living area.[20]

The plants in the Villa Tugendhat served many purposes, therefore. They reinforced the formal and spatial strategies of the architect; they helped the inhabitants to negotiate the spaces; they pinpointed and resolved points of spatial tension; they demarcated boundaries, however permeable; they contributed to the creation of inside/outside ambiguity; and they emphasised the villa's having been designed from the inside out. They were also one of the multiple softening agents used to transform that potentially hard and dehumanised building into a dwelling, even into a home, albeit a modern one; they helped to maintain some continuity with the past; and they offered the possibility of well-being. Finally, they contributed to a sense of the exotic that had distinguished the interiors of wealthy inhabitants for decades.

While the Villa Tugendhat contained a highly progressive, modernist space rather than a Victorian interior, nature inside helped ease the transition from one to the other. The result was a new model of modern domesticity. In the years after 1945, several European émigré architects took that model across the Atlantic and influenced the indoor spaces of many of the new residences constructed on the west coast of the United States.

CHAPTER 6

Living in the Garden: Californian Modernism

We're terribly House and Garden,
Now at last we've got the chance.
The garden's full of furniture . . .
And the house is full of plants.[1]
Flanders and Swann

When European architectural modernism spread to the west coast of the United States in the 1920s, it encountered a climate that was warm and dry all year round, with lush vegetation. While the programme of social utopianism, and the alignment with cultural and technological modernity remained in place, many of the private residences that were created in this new setting were heavily influenced by their encounter with nature. As a result of the indoor/outdoor lifestyle made possible by the weather, it became increasingly difficult to differentiate between the two spheres. What constituted inside became less clearly defined, and the close relationship between the garden and the house that Humphry Repton had initiated back in the eighteenth century was dramatically intensified. As a result, a discussion about nature inside in this context becomes, by necessity, a more open one, embracing what, in another context, would have been more clearly the space of the garden.

Modernism was brought to the west coast of the United States by several European émigrés seeking a new life. Among them were the Viennese architects Rudolf Schindler and Richard Neutra. Schindler arrived in southern California in 1921 to work on Frank Lloyd Wright's Hollyhock House for Louise Aline Barnsdall (fig. 64). Neutra, a friend of Schindler's from 1913, also worked for Wright for a short period but, following an invitation from Schindler, moved to the West Coast in 1925. Neutra's first projects, which were in landscape design, included a garden for a beach

Fig. 64 Ferns in a built-in planter in Frank Lloyd Wright's Hollyhock House, Los Angeles, completed 1921. Photograph by the author, 2017

house designed by Schindler for Philip Lovell, and a pergola and wading pool for the Barnsdall house. The Swiss architect Albert Frey, who had previously worked in the office of Le Corbusier, was among the numerous other Europeans who moved to southern California. He arrived in the United States in 1928 and settled in Palm Springs.

With the possibility in California of not merely looking at nature outside through plate-glass picture windows (as at the Villa Tugendhat), but rather of actually living outside for much of the year, and, in so doing, taking the inside outside, nature inside took on a new meaning. The model was the Californian ranch house, with its space for outdoor entertaining.[2] The patio and the garden became outside rooms, the former often furnished almost as if it were inside. In that new domestic scenario, the ways in which plants and flowers were used became highly strategic and the idea of landscape, no longer restricted to the garden area but also part of the inside spaces of the house, became an important component of a new, late-modern language of domesticity.

With its easy access to mountains, deserts and the shoreline, 1920s southern California has been described as 'a bucolic Eden'.[3] A modernist architectural movement developed there that was 'looser, warmer, more ad hoc' than its European equivalent.⁴ Not only did the Californian climate facilitate outdoor living, it also made it easy to embrace the idea of the interior open plan, which could now flow easily into outdoor spaces. That strategy had already revealed itself on American soil earlier in the century, in the work of Frank Lloyd Wright and his Prairie School colleagues, and in that of members of the American Arts and Crafts movement, including Charles and Henry Greene, whose Gamble House was described in Chapter 4. Following Greene & Greene, Schindler picked up the idea of integrating open sleeping porches into a house when designing a residence in Kings Road, Los Angeles, which was completed in 1922. The house was built for two couples to live together, sharing a kitchen. Heavily influenced by Japan – as Greene & Greene had been – the outside was welcomed in (and vice versa) through the use of fragile sliding screens made of gypsum board in the rooms that opened onto the rear garden. In turn, the garden was enclosed on three sides by the house and acted as an external room for most of the year. Outdoor fireplaces made that an even more feasible proposition.

Neutra also negotiated the relationships between the insides and outsides of his houses, and cultivated their rapport with nature. Having worked on the Demonstration Health House in Los Angeles (1927–9) for the naturopathic doctor Philip Lovell, who had already employed Schindler, he established his own practice in 1931. Featuring the first use of a steel frame in a domestic project, the Lovell House was a material representation of the doctor's idea that open-air sleeping was linked to healthy living. Vast expanses of glass were used to bring the outside in, and sleeping porches and patios were added to 'extend living outwards'.⁵ 'We love nature, plants, views, light and air, differently from, and more than, previous generations', Neutra proclaimed, in direct reference to plants particularising the generalised view of nature that had been embraced by the inter-war European architects.⁶

Thus, when Neutra was commissioned by Grace Miller to build a small house in Palm Springs, which was completed in 1937, his ideas about health, nature and the therapeutic role of architecture were already fully formed (fig. 65). As the house was built in the desert, with a view of mountains, nature played a key role in its conception. As well as working with plants and flowers, Neutra understood nature in an abstract sense as the laws of biology that determine the movements of the body. Miller wanted to use her home as a studio in which to teach the 'Mensendieck System of Functional Exercises', and Neutra was fully aligned with the project from the outset.⁷ His views were rooted in what seemed like a fundamental contradiction – that the body, like architecture, was a machine but that it was also rooted in nature. The resolution of that apparent contradiction lay in his belief in the union of the pastoral ideal and the machine culture, a belief that underpinned American modernism in general. In his 1964 book *The Machine in the Garden: Technology and the Pastoral Ideal in America*, the historian of technology Leo Marx was to claim that this union characterised the culture of the United States after 1945.⁸

Fig. 65 Interior of the Grace Miller House, Palm Springs, CA, designed by Richard Neutra, completed and photographed 1937. Photograph by Julius Shulman

In order to integrate it into the landscape, Neutra ensured that the Miller House opened directly onto the desert. A screened patio with a reflecting pool reinforced that link. Miller explained that 'Mr. Neutra was given a clear idea of my preference for taking meals outdoors.'[9] The living space was left open, leading onto the studio and the patio. Neutra described the house as 'nature near'.[10] The landscaping of the garden was also carefully considered by the architect, who planted cacti and desert foliage near to the house to replace the planting that had been lost during its construction. In the garden area he created both 'an idealized desert setting and an oasis', representing, in his words, 'both nature and culture'.[11] To create his desired effect Neutra combined a number of non-indigenous, exotic cacti with native trees and shrubs – among them a smoke tree, a cottonwood tree, desert willows, a mesquite, a range of citrus trees, greasewood bushes, a white oleander, desert verbena and tamarix shrubs. He used a stone walkway to separate the distinct areas.[12] Neutra linked the garden to the inside décor by using colours, including rust, taken from the desert plants in the garden.

The photographer Julius Shulman, who was based on the West Coast, created several images of the Miller House.[13] Writing about a group of six of them that appeared in the May 1937 issue of *Architectural Record*, Stephen Leet has explained that:

> Neutra, Shulman and Miller cleverly rearranged the few furnishings she [Miller] had – a glass-topped coffee table, two Navajo saddle blankets, a rattan chair, a potted manzanita tree, flower vases and metal bowls, a few magazines and books and a Neutra-designed floor lamp – recomposing each room with essentially the same pieces to create six interior views.[14]

Four sets of Shulman photographs of the Miller house were produced between 1937 and 1941. Vases of flowers are often visible, undoubtedly placed there for photographic reasons, to soften the hard lines of the architecture, to create focal points and to help articulate spatial relationships.

Writing about Neutra's 1946 Desert House, created for Edgar J. Kaufmann in Palm Springs, the architectural historian Alice T. Friedman has pointed out that the architect exercised strict control over how the interior was photographed. 'Neutra', she wrote:

> laid out his instructions for photographing the interior of the Kaufmann house. He explained that, 'the particular house for which Mr Kaufmann, as myself, has made great personal effort and which has subtle aesthetic qualities can be taken like for a high-class photo-magazine rather than for a furniture review. Thus, when in doubt, eliminate the furniture, or even the picture from its place; we may have in mind to reconsider the position.'[15]

Fig. 66 Grace Miller in the porch of her house, Palm Springs, CA, designed by Richard Neutra, completed and photographed 1937. Photograph by Julius Shulman

In the Miller House vases of flowers were positioned, for example, on some low shelving and on a round coffee table in the living room, as well as on the floor in the studio. In the widely reproduced image of Grace Miller sitting in the glazed porch leading from the living room, reading a book, with a view of the desert and the mountains in the distance, a metal bowl placed on the floor and containing flowers (which, from the archive images, look like tree blossoms and carnations) creates a boundary where the porch joins the pool, thereby emphasising the continuity between inside and outside (fig. 66). It also acts as a counterpoint for the tamed potted plants at the other side of the pool, and for the wild desert plants beyond them, making a subtle commentary on the transition from tamed to wild nature, a strategy which, inspired by Mies van der Rohe and others, had, by that time, become a familiar modernist trope. It is notable that the bowl is situated at the very point at which the sunlight turned to shadow on the floor when Shulman was taking his photograph. Forty years later, Shulman explained that 'the true landscape photograph is one in which not only the plant material is depicted, but the setting', adding that 'it is important to constantly relate the garden to the house'. He also maintained that he was interested in shots in which 'the sun uniformly illuminates the interior–exterior spatial volumes that show the blend of architecture, interiors, and landscape in a purposeful way'.[16] Those ideas were clearly already informing his approach to taking photographs of houses in the 1930s.

The optimism, democratisation and technological progress that had characterised architectural developments in southern California before the Second World War continued into the post-war years, which also witnessed the expansion of the film industry and the emergence of a range of glossy magazines that took the utopian vision to a wider public. In addition, war-time technological breakthroughs – such as the ability to make larger glass panels, and developments in the production of steel, used for sliding doors – made the ideas developed by Neutra and others accessible to greater numbers of people.

Partly as a result of population growth and partly to house the returning military, there was a critical need for affordable new housing in the post-war years. Projects such as the Mutual Housing Association (MHA), on which the architects A. Quincy Jones and Whitney R. Smith, among others, worked, were established to address that need. Of course, not all new post-war housing in southern California was designed in the modern style. The dominant idioms were Spanish, ranch style, Colonial and Victorian, among others. However, one Californian architectural project – the Case Study Houses, initiated in 1945 by John Entenza in his role as editor of the journal *Arts & Architecture* – focused on the possibility of creating low-cost prototypes with modular standardised parts fabricated from modern materials in the modern style.[17] Thirty-six designs for two-bedroomed, single-storey houses for people without servants were commissioned from notable architects of the day (some of them were never built). One of the most notable features of many of the built Case Study Houses was their relationship with nature. They reinforced, yet again, the American dream of bringing together the pastoral ideal and the technological future.

Case Study House 8 in Los Angeles, designed by Charles and Ray Eames as their own home, and completed in 1949, demonstrated a close relationship with nature in a number of ways. Originally designed in 1945 as a cantilevered structure, its 1949 redesign was the result of a decision to site the house in a small meadow at the foot of a hill, behind a row of eucalyptus trees. Shulman's 1949 photograph of the house shows the trees reflected in the colourful, standardised, geometrical units of the house, softening their hard lines, with the trunks of the trees mirroring the house's supporting pillars. A row of small pots of plants and flowers positioned just outside the glass wall of the building made a link to nature outside, both marking and blurring the boundary between house and meadow (fig. 67). Over the years the pots grew in number. Gradually gardens appeared at the front of the main house and in the space between it and the studio. Geraniums had a fairly constant presence. In a photograph Shulman took in 1950, a small potted geranium had infiltrated the interior, carefully positioned on the floor near a banquette

Fig. 67 Potted plants mask the boundary between house and garden, Eames House, Los Angeles, designed by Charles and Ray Eames, 1949. Photograph by the author, 1992

in an alcove (fig. 68). It was accompanied by a large rubber plant that marked the end of the banquette, and a partly visible Swiss cheese plant used to frame the image. Although Shulman sometimes had people hold plants up as a means of creating frames for his photographs, and to give them depth, in this instance he simply used the plants that were already in situ to frame his image. Rubber and Swiss cheese plants had been popular since the nineteenth century. With its shiny, leathery leaves the rubber plant is less obviously beautiful than the palm but it has its own dignity nonetheless. It is easy to cultivate and content with shade. Swiss cheese plants also have shiny, leathery leaves, whose dramatic slits give the plant a sculptural appearance in an interior setting.

Another of Shulman's 1950 photographs was taken from the space between the two buildings, looking directly at the pillar that demarcated a corner of the house, and that divided the image in two. In a single shot Shulman was able to capture the trees outside, the small potted plants at the boundary between the house and the meadow, and the larger indoor exotic plants, thereby collapsing the interior and the exterior into a single space (fig. 69). The combination of indigenous trees, potted outdoor plants

Fig. 68 Potted plants inside the Eames House, Los Angeles, designed by Charles and Ray Eames, 1949. Photograph by Julius Shulman, 1950

Fig. 69 Looking into the Eames House, Los Angeles, designed by Charles and Ray Eames, 1949. Photograph by Julius Shulman, 1950

and exotic houseplants represents the spectrum of nature, from tamed to wild, that the Eameses used to embrace both a local sense of place and a generalised model of nineteenth-century domesticity that depended upon the inclusion of an exoticised 'other'. Another of Shulman's photographs from 1950, taken from inside the house looking out, cleverly uses the indoor plants to create an effect of double framing (fig. 70). The leaves of a Swiss cheese plant frame the image, while a rubber plant, positioned at the far side of the room, frames the view from the interior onto the outside space beyond it.

The Eameses had already used indoor plants in their flat in the Strathmore Apartments, designed by Neutra in 1941. A photograph of the couple in that space depicts them surrounded by rubber and Swiss cheese plants, together with a large, abstract, biomorphic image on the wall, the shape of which resembles those of the leaves.[18] The Eameses' plants – probably the same ones – moved with them into Case Study House 8 and were positioned strategically to help articulate the distinct areas within that open space. Over the years, increasing numbers of personal objects were also introduced, and the plants continued to grow. Photographs from the 1960s show the nature inside the house looking increasingly wild. An image from 1961 depicts a large Swiss cheese plant; by 1966 it had taken

Fig. 70 Indoor foliage frames an interior shot of the Eames House, Los Angeles, designed by Charles and Ray Eames, 1949. Photograph by Julius Shulman, 1950

Fig. 71 Large indoor plants growing in the Sheats Goldstein House, Los Angeles, designed by John Lautner, completed 1963. Photograph by Gaelle Le Boulicaut, 2010

over an area of the house. A later colour photograph, taken in 1974, shows that it grew significantly in the intervening years, nearly reaching the ceiling. In that same image, vases containing small posies of colourful flowers, arranged by Ray, can be seen on the floor. The image therefore demonstrates clearly that Ray's 'functioning decoration' (to borrow Pat Kirkham's term) was in place in the house by that date.[19]

As can be seen in one of Shulman's photographs from 1958, Charles also introduced a number of potted plants into his studio, among them a small Australian umbrella plant. Other images showed air plants attached to the wall of his room. The exoticism of the plants in the Eameses' house matched that of their many souvenirs brought back from Mexico, India and elsewhere. Although supremely modern, with its strong references to exoticism, its clutter and, above all, its inclusion of nature, Case Study House 8 also clearly embraced a model of domesticity that had been formed in the nineteenth century.

The English architects Alison and Peter Smithson, who admired the Eameses, and who designed a small 'pavilion' in Wiltshire, England, in emulation of Case Study House 8, understood the role played by nature in their Los Angeles model. They explained that a pavilion is 'a place wherein to be restored to oneself; as a source of one's energies. The pavilion is thus seen as a place made idyll; a dream of a stress-free way of life, a domain – often a greater garden – often in the pretend wild; that is, nature.'[20]

Nature and plants were used in a variety of ways in several other southern Californian mid-century modern houses. First and foremost, like the Eames House, many of them were constructed on sites covered in natural vegetation, whether in areas that featured tropical trees, plants or flowers, or in the desert, where a different kind of indigenous vegetation could be found, including cacti and succulents. Some of them, for example A. Quincy Jones's MHA site office of 1947, which was later turned into a home, and John Lautner's Sheats Goldstein House of 1963 (fig. 71), were enveloped in luxuriant greenery. Others, for instance Albert Frey's House II of 1964, which featured an internal rock that acted as a partition between the living and the sleeping areas, and Lautner's 1962 Garcia House, were constructed on the sides of hills or mountains. Their views into the distance were their defining features. Yet others, among them Harry Gesner's Wave and Sandcastle houses of 1957 and 1970, built on the beach at Malibu, looked out across the ocean while also being partly enclosed in tropical vegetation.

Those houses, and many others like them, were defined by the nature that surrounded them and were designed to exploit the sense of drama and beauty that it bestowed. Arguably, that continuing respect for nature served to keep mid-century modernism from over-reaching itself and losing touch with its buildings' human inhabitants. For those people who made the brave choice to live in machine-age houses, plants helped to alleviate the anxiety they might otherwise have experienced in removing themselves from everything that was familiar.

Fig. 72 Plants growing in an internal flower bed in Case Study House 1950, Los Angeles, designed by Raphael Soriano, completed and photographed 1950. Photograph by Julius Shulman

Sometimes nature was planted directly into soil beneath the floor, thereby recreating natural elements that had existed before the buildings obliterated them. That practice was evident, for example, in Raphael Soriano's 1950 Case Study House in Pacific Palisades, in which the architect planted a group of potted palms and Swiss cheese plants, among other exotics, in an internal square bed situated beneath an opening in the ceiling (fig. 72). Neither fully inside nor outside, the plants served to reinforce the ambiguity and contradiction that defined so many of these houses. Case Study House 16 in Beverly Hills (1947), designed by Rodney A. Walker, also featured an indoor bed containing a variety of plants. Positioned immediately next to floor-to-ceiling glazing, it was the continuation of a bed of plants located on the shielded exterior patio that extended beneath the lower steel frame of the windows, thereby creating a close link between the exterior and the interior.

A. Quincy Jones also had a predilection for indoor landscaping. In his Brody House of 1949, on which he collaborated with the landscape architect Garrett Eckbo – who worked on several Californian modern houses – and the interior designer William Haines, the architect featured, among numerous pots and troughs of indoor plants, a large indoor tree. In his 1965 interior refurbishment of the Barn, Los Angeles, Jones placed plants directly into beds in the floor. In 1956 Eckbo encouraged his readers to leave trees in place when constructing a patio.[21] The ultimate aim, he explained, was to disturb nature as little as possible.

Figs 73 (*left*) and 74 (*right*) The planting scheme creates a sense of permeability between outside and inside, and a boundary line behind a sofa at Edris House, Palm Springs, CA, designed by E. Stewart Williams, completed 1954. Photograph by the author, 2017

In his Edris House of 1954 in Palm Springs, E. Stewart Williams created an indoor bed of mother-in-law's tongues right next to the glazed front door, which was mirrored by an external bed of different plants placed just outside the door and visible through the glazing. Both plants were embedded in grey pebbles (fig. 73). While continuity between inside and outside was thus maintained, the use of different plants meant that it was simultaneously eroded. Elsewhere in the house, Williams planted a row of mother-in-law's tongues in a bed immediately behind a sofa in the living area, thereby creating a screen that separated the sofa from the walkway behind it (fig. 74). Similarly, plants were placed at the meeting point between a built-in shelving system and a fitted sofa in Charles Eames and Eero Saarinen's Case Study House 9 in the Pacific Palisades (1949). In the open-plan settings of many mid-century modern houses, plants were placed at the boundaries formed by bookcases or room dividers, or at the bottoms of flights of stairs, helping to mediate spatial intersections and transitions. Indoor and adjacent outdoor plants were used to define and soften corners, levels and other spatial articulations.

The use of the same ceramic plant containers in both the indoor and the outdoor spaces of many mid-century Californian modern houses provided yet another means of eroding the boundaries between the two spheres. The firm Architectural Pottery, formed in 1950 by Max and Rita Lawrence, supplied a range of ceramic pots and planters that could be

Living in the Garden

used both inside and outside.[22] The firm was initially established to produce and sell the pots that LaGardo Tackett created with his students at the California School of Art.[23] The Lawrences saw a market in the architects and owners of the new houses, and they employed John Follis, who also designed issues of *Arts & Architecture*, and Rex Goode as designers. In 1957 wooden stands were introduced to integrate the planters with 'the softened, elegant, carpeted living-dining-entry room décor'.[24] In their own garden the Lawrences constructed a tall pile of their conical planters that resembled a Brancusi sculpture. Others used their products for a range of purposes, from screens to pipe concealers to fountains.[25]

The stark, architectural profiles and plain colours (mostly black and white) of Architectural Pottery's products provided dramatic punctuation marks inside and outside houses and at the boundaries between the two. They also reinforced the visual contrast between geometric abstraction and the fluid, ill-defined, often chaotic shapes and profiles of the greenery they contained. In houses where the rooms opened onto patios constructed on the same level, they confirmed that important continuity, adding to the sense of ambiguity and flow. Acting as space and boundary definers, as way-finders, and as concealers of features that architects wanted to downplay, they quickly became familiar features of 1950s Californian modern homes.

In his 1958 Bailey House (Case Study House 21) Pierre Koenig used Architectural Pottery planters to help define boundaries and spatial relationships, to resolve differences and to ease contradictions. Koenig had designed his first steel-framed house for himself and his family in 1950. Two years later he was commissioned by Walter Bailey, a psychologist, and his wife, Mary, to design a house in Hollywood Hills Canyon. He created ponds around the house to make it look as if it were floating. The reflections of the trees helped soften the building's minimal form and its materials.

Koenig used potted plants to the same effect. Outside, on the north-facing side of Case Study House 21, a large, white, conical Architectural Pottery pot, filled with a big jade plant and other greenery, and raised from the ground on a wooden frame (not used inside on that occasion), was used to emphasise the point at which the carport finished and the entrance porch began. That same boundary was also demarcated by a change in the material used for the flooring, from grey macadam to pink brick. The addition of the potted plant mediated and softened what would otherwise have been an abrupt transition, and permitted a smooth entrance to the house. It also marked the boundary between the carport and the pond that ran alongside it (fig. 75). Two more white conical pots, mounted on steel rods and containing fatsias, were placed in another pond immediately in front of the parked car. A small Saarinen-designed table was often brought out onto that section of the patio for outdoor dining. The indoor/outdoor feeling thus created was reinforced by the presence of a tall, slim, potted tree that also sat in the water and acted as a screen for the diners, helping to conceal the parked automobile while they were eating.

Fig. 75 Potted plants help articulate the indoor–outdoor continuum at the Bailey House (Case Study House 21), Los Angeles, designed by Pierre Koenig, completed 1958. Photograph by Julius Shulman, 1958/9

Koenig's desire to minimise the distinction between the outside and inside spaces of his building was further realised through the inclusion of yet more potted plants, contained in similar planters, positioned within the house itself. A number of them, including another two white conical pots mounted on steel rods, were placed in a row in a built-in, floor-level trough that ran alongside a low partition wall dividing the entrance area from the main living space. In addition, a black, waisted pot containing a tall asparagus fern that almost reached the ceiling was placed on a credenza immediately opposite the trough, creating a dramatic punctuation mark. A potted rubber plant was placed in the rear corner of the living area, while yet more potted plants were used to demarcate edges, corners and boundaries and to mitigate the effects of the material variations employed on the south side of the house.

As an advertisement for the firm Roy F. Wilcox and Company, based in Montebello, which described itself as 'the west's largest wholesale growers of decorative plants', explained: 'Proper planned interior plantings afford the architect countless opportunities for harmonious transitions from traditional to modern.'[26] The same company also explained that plants helped to create

Living in the Garden 121

moods in houses. Like the desert sun, they were also considered to promote health and well-being. Spatially, they helped to link pieces of furniture to floors and walls, and they worked with the furniture arrangements to encourage casual living and relaxation. Some coffee tables even had special compartments built into them to hold plants.

It is not always possible to know who decided to include the plants that featured so prominently in the interior and almost-interior spaces of so many Californian modern houses. In the Eames House, Charles and Ray (more probably the latter) clearly chose to have plants as living companions; in other spaces it could have been the architect, the landscape designer, or the inhabitant – or a combination of all three – who selected them and chose where to place them. In some instances, they may have been added as props by photographers who wanted to soften the otherwise hard geometry of the houses, to make their images more picturesque, to frame them and to provide a sense of depth.

By the post-war years the glamour of Hollywood had begun to rub off on many other areas of life in southern California, including the images of modern houses that were depicted in popular magazines of the day – *House Beautiful*, *Ladies' Home Journal* and *Life* among them. The exotic, luxuriant plants that were often used added to that glamorous effect. Shulman was a master at making Californian modern houses look not merely glamorous, but even dangerous. Nowhere is that more evident than in his famous image of Case Study House 22 (1950, designed by Koenig), which is cantilevered precariously over the city of Los Angeles. The people depicted in the house were, as so often in Shulman's photographs, and like many of the plants, models brought in for the shoot. The rubber plant partially visible in the left of the image plays a key role in making the cantilevered room appear to project far into the distance.

The professional distinctions between the people who created the interiors and exteriors of the Case Study Houses were often unclear. Garrett Eckbo acknowledged that 'in addition to providing beautiful settings for the house, and lovely pictures to view out of the windows, garden spaces influence indoor comfort, and during the milder portions of the year can be developed into outdoor extensions of the living spaces of the house'.[27] And just as architects and landscape architects overlapped, so too did the suppliers of plants for inside and those who supplied plants for outside use. Roy F. Wilcox had established his nursery in the 1920s when the trend for interior landscaping had first taken off, and he seems to have developed a niche market for his products. A 1949 advertisement for the company, placed in *Arts & Architecture*, which was clearly targeting architects working in the Californian modern style, featured a range of decorative foliage plants, including the familiar Swiss cheese plant, and promised 'professional discounts for architects, designers and decorators'.[28] It also offered to help those professionals interpret how to deliver what they wanted through plants. The firm continued to put adverts in the same publication over the next couple of years, although they were gradually reduced in size. The firm closed in the mid-1950s, probably as a result of competition from larger nurseries that began supplying a wide range of plants and flowers.

Although indigenous trees, plants and flowers were widely used in Californian modern houses, they were frequently supplemented by exotic, non-indigenous imports. The oft-cited idea that these buildings were linked to a sense of place through their use of indigenous vegetation is, therefore, a little wide of the mark. In his library in the Barn, A. Quincy Jones kept a colourful and well-thumbed book about exotic plants. The Swiss cheese plants, rubber plants, bamboo, and many of the cacti and succulents that were used in Californian modern houses originated outside California. Yet as they sat alongside indigenous trees, grasses and cacti, the difference between them was as cleverly eroded as was that between the inside and the outside of the buildings themselves.

The fact that the outside areas of many Californian modern houses were designed as rooms was frequently reinforced by the accompanying furniture. Sometimes, as was the case with wooden benches, often made of Californian redwood, they were designed specifically for the garden. Sometimes they employed the same materials – fabric upholstery, for example – as could be found indoors. Several attempts were made to design furniture items that could be used both inside and out, many of which were made of metal or woven rattan. Rattan was considered an exotic material and chairs made from it were often positioned among tropical planting. Charles Eames designed a range of aluminium furniture for outside use, which could also be used indoors. Eckbo was responsible for a number of outside spaces that contained elegant wooden benches, while examples of indoor fabrics on outdoor seating include those utilised on the built-in seats that Albert Frey added to two sides of his second residence in Palm Springs, and the upholstery added to the wooden seating that A. Quincy Jones designed for his garden at the Barn.

The new model of modern domesticity that emerged in post-war California, which offered people an inside/outside experience, was quickly emulated across the United States, especially in the new residential suburbs where, in a popularised form, alongside bulbous automobiles and refrigerators, it became a component of the 'American dream'. In contrast, through the 1960s many of America's urban centres witnessed the emergence of large late-modern developments, often high-rise, constructed to meet the growing needs of commerce, leisure and work. Those buildings also embraced indoor planting schemes, introduced to humanise their otherwise harsh and impersonal spaces.

CHAPTER 7

Natural Late Modernism

Living among plants has a humanizing effect and helps to restore to a healthy condition minds that are burdened with grief, pain and trouble.[1]
George H. Manaker

While in 1950s California, a new model of domestic modernism offered people an indoor/outdoor modern lifestyle in which the natural world played a key role, the urban areas of the United States were being transformed by an expansion of office work, the commercial ambitions of large corporations, and a growth in consumption and public leisure.[2] To meet the needs of the new spaces in which those activities took place, and because large sums of money were being spent on them, the American interior design press advocated a shift in designers' activities from residential to non-residential work. As Harvey Anderson explained in the April 1955 issue of *Interior Design* magazine, 'It is the interior designer with experience in residential work who is doing the best job in the contract field', adding that 'he has taken the commercial stigma out of non-residential interiors by giving to them a distinctly residential character.'[3] Nature inside played an important role in that transfer of 'residential character', or domesticity, into the public sphere. As a means of making people feel at home and comfortable, greenery began to appear in numerous offices, hotels, restaurants, banks, shops and shopping malls. Given that their meanings were rooted in the anti-commercialism of the domestic interior, plants were able to perform that task very effectively.

As was demonstrated in Chapter 3, in the second half of the nineteenth century winter gardens, people's palaces, department stores, grand hotels and department stores had been transformed by the addition of exotic plants that generated an ambiance of leisure and luxury. The United States' post-1945 inside spaces of consumption and leisure had their roots in that earlier era and, like their antecedents, they offered antidotes to the numerous ills that accompanied urban life. Not until the 1940s, however, did a group of specialist design professionals emerge to make all this possible. While in earlier decades indoor planting had been undertaken by architects working directly with commercial nurseries, by the 1940s architects were frequently working with interior decorators and designers, landscape architects, interior landscapers (increasingly also referred to as interiorscapers and plantscapers) and horticulturists to create plant-filled interiors.

The mid-century desire in the United States to (re)turn to the natural world was noted in 1964 by Leo Marx. 'An inchoate longing for a more "natural" environment enters into the contemptuous attitude that many Americans adopt towards urban life', he wrote.[4] In support of that 'longing', exotic plants and flowers, and often trees, were brought into a range of new urban commercial buildings that supported dining, shopping and other leisure and commercial activities. The idea of Paradise on earth also resurfaced in that new context.

The new commercial and corporate spaces were mostly privately financed and their owners could, more or less, select those who entered them. Consumers were invited into a world of affluence and luxury in which they were made to feel at home. In addition, many of these commercial buildings sought to deny their urban-ness by turning inwards, ignoring the environments outside and creating cocooned inside spaces.[5] The introduction of nature into many late-modern indoor urban environments, which would otherwise have been exclusively defined by their commercial agendas, was intended to provide a sense of harmony and identity. It also brought texture and colour to compensate for the buildings' spare forms, industrial materials and, often, superhuman scale. Bringing nature inside also introduced a sense of the changing seasons into those unnatural, man-made, static structures.

This chapter explores three examples of American plant-filled interiors in late-modern commercial and corporate architectural structures: New York's Four Seasons Restaurant of 1959; Dallas's NorthPark Center of 1965; and New York's Ford Foundation Building, an office complex completed in 1968. All three had plants and flowers introduced into them, interventions which required their architects and interior designers to collaborate with a range of other professionals. The Four Seasons Restaurant in New York, which was designed from the outset to contain trees and a range of plants, was the result of input from two architects, an interior designer/decorator, a landscape architect and an interior landscaper/horticulturist. The restaurant group that financed the project also influenced the design decisions, including the introduction of plants.[6]

When Ludwig Mies van der Rohe completed the Seagram Building on Park Avenue and 52nd Street in 1958, it was clear that something was needed to fill a space in the ground-floor lobby area and to attract visitors

Fig. 76 (*from left to right*) Philip Johnson, Ludwig Mies van der Rohe and Phyllis Lambert at the Museum of Modern Art, New York, c.1955. Photographer unknown

to the building. Conversations took place between Phyllis Lambert, the daughter of the CEO and Head of Planning at Seagram, and Mies as to what that should be (fig. 76). A museum, an antiques shop, a bank showroom and a car showroom were all discussed but, in the end, Jerry Brody, an executive of a company called Restaurant Associates that was running the project, suggested a restaurant, specifically one that would appeal to a modern, discriminating and distinctive clientele. Philip Johnson, the director of the Department of Architecture at New York's Museum of Modern Art (MoMA) at the time, was tasked with designing it, or at least with coordinating its design.[7]

In July 1959 the restaurant opened to the public. It contained a Grill Room and a Pool Room, the latter earning its name from the 20-foot-square Carrera marble pool positioned at its centre. As the Pool Room was a large space, it needed something other than tables and chairs to fill it. A decision was made to add four trees at the corners of the pool, as well as some potted plants suspended from a horizontal rail that spanned the floor-to-ceiling windows down one side of the room (fig. 77). The trees were planted in large round pots with flowers growing below them. As the room had

Natural Late Modernism

Fig. 77 The Four Seasons Restaurant, New York, 1959. Photographer unknown

20-foot ceilings, trees 17 feet tall were used to make the necessary visual impact and to fit with the scale. Originally, four fig trees were introduced. Later azaleas and birch trees were used in the spring; philodendrons and queen palm trees in the summer; burnt-orange and yellow chrysanthemums and oak-leaf branches in the autumn; and white chrysanthemums and white birch trees in the winter. The planting scheme changed over the life of the restaurant and, at one point, blossoming cherry trees made an appearance.

Before Restaurant Associates, which was co-run by Jerry Brody and Joe Baum, made their pitch for the Four Seasons, they had worked on a number of other smart, themed restaurants, including the Hawaiian Room in the Hotel Lexington, which was appropriately decorated with palm trees and exotic plants. The Forum of the Twelve Caesars in the Rockefeller Center, on which they had worked with the interior decorator William Pahlmann, was another of their restaurant projects.[8] In 2009 Phyllis Lambert recalled that, 'Philip and I went with Jerry Brody out to New Jersey and had a meal at his restaurant, where they gave us soup in profiterole bowls. Then we went to the completely over-the-top Forum of the Twelve Caesars, designed by Bill Pahlmann' (fig. 78).

Fig. 78 William Pahlmann looking through pattern books with Mrs Walter Hoving, one of his clients, 1948. Photograph by Horst P. Horst

Brody and Baum had established the typology for luxurious, highly decorated, themed restaurants, the aesthetics of which were a far cry from the restrained, minimalist, yet also highly luxurious, approach advocated by Johnson for Mies's Seagram Building. But it appears that Brody and Baum had significant influence over the designs of the restaurants they worked on, and the Four Seasons was no exception. They undoubtedly suggested that Pahlmann, whom they had used on several of their projects, be brought in to help with the restaurant's interior design. As John Mariani and Alex von Binder later explained, Lambert and Johnson 'were convinced the arrangement [with Pahlmann] was not going to go smoothly as they felt that they were diametrically opposed in their attitudes'.[9] It seems that Johnson and Pahlmann did dislike each other but, at a later date, Johnson was to claim that 'Pahlmann was very very helpful with the interior, especially with the table placement and kitchen layout'.[10] The pool and the use of chairs based on Mies's Brno design in the restaurant were apparently Pahlmann's ideas.

Pahlmann had made his reputation in the inter-war years and the 1940s as a decorator of expensive homes, and his interiors were highly eclectic, combining traditional and modern styles. His move to the contract field

Natural Late Modernism   129

– restaurants included – was in line with the comments made in the April 1955 issue of *Interior Design*. His contributions, however minor, undoubtedly helped to bring a level of domestic comfort and luxury into the Four Seasons.

The realisation of an interior scheme that involved nature required yet another kind of expertise, however. Although it was an interior project, the scheme for the Four Seasons required the knowledge base of someone who was used to working with plants and landscapes. The landscape architect Karl Linn (who, in later life, was a key advocate of using shared gardens to create communities) was employed to meet that need.[11] Arguably, this is an early example of what later, first in the office context and subsequently more widely, became known as an 'interior landscape', a point that was recognised by Mariani and von Binder when they wrote: 'Here was a restaurant whose interior was actually *landscaped*, not just set with flowers and potted plants.'[12]

Linn, who had already designed the exterior green spaces of the Seagram Building, was commissioned to create an interior design with plantings keyed to the theme of the four seasons. He rejected the idea of Picasso sculptures round the pool, and Restaurant Associates decided that trees should be used instead to serve as a kind of sheltering canopy. The project required Linn to undertake a great deal of research, and he visited many nurseries to find matching trees, which had to be pruned before shipment. With the plant psychologist O. Wesley Davidson and the lighting expert Richard Kelly, Linn developed a system that provided consistent humidity, as well as lighting and temperature controls that would allow for appropriate maintenance.

Inspired by the fact that the trees were to be replaced four times a year, the idea of seasonality became the restaurant's theme (and name). Baum admired two existing Four Seasons hotels in Munich and Hamburg and, following a suggestion from the editor of *House Beautiful*, Elizabeth Gordon, decided that 'everything from plants to ashtrays would be chosen and designed to fit the idea of passing seasons'.[13] Even the staff uniforms and the upholstery changed four times a year. The strategy also fitted with the idea of the food being contemporary and in tune with the seasons. In the form of seasonally rotated trees and plants, and, more subtly, as part of a commitment to the very idea of renewal, nature was introduced to create a sense of change in an otherwise static late-modern environment. It softened the pervasive monumental aesthetic and added a human dimension. Just as Johnson and Pahlmann needed to engage Linn as a consultant, so the landscape architect was, in turn, dependent on the expertise of Everett Lawson Conklin, the horticulturist and interior landscape expert employed to work on the project. It was he who advised on the trees and plants, supplied them from his nursery, and helped maintain them.

The interior landscape industry developed in earnest in the late 1940s, though there was still only a small body of practitioners – Conklin among them. Describing himself early on as an 'interior planting specialist', his experience as a sales manager and division manager for the Bobbink & Atkins nursery in East Rutherford, between 1938 and 1957, had prepared him for his future career in horticulture. From 1957 until his retirement in 1982, Conklin was the president and chairman of Everett Conklin & Company International of Montvale, New Jersey. In addition to his input into New York's Four Seasons Restaurant, he later worked on the Philharmonic Cafe at the Lincoln Center, the Promenade Cafe in the Rockefeller Center and the indoor garden in the Ford Foundation Building. He also created the dramatic plantscaping in the Crown Center Hotel in Kansas, and was the chairman of the floral decorations committee for President Richard M. Nixon's inaugural balls in 1969 and 1973.[14] One of Conklin's key contributions to indoor horticulture was his discovery of the need to acclimatise plants to low light levels before installation. By the early 1970s his nursery business had become one of the largest interiorscaping companies in the United States, specialising in shopping malls and hotels.

As it had been in the nineteenth-century domestic parlour, and in the Villa Tugendhat, nature in the Four Seasons was introduced as a form of art, albeit now as a form of modern art that helped create a space of contemporary affluence and taste. Plants and flowers sat comfortably alongside the Richard Lippold sculpture located above the bar in the Grill Room, the metal chain drapes designed by Marie Nichols, and the artworks by Picasso, Miró and others. The works of art in the restaurant had been approved by MoMA's fine-art curator, Alfred H. Barr. The furniture and designed objects in the restaurant – among them the banquettes designed by Johnson, the Eames private-party chairs, Eero Saarinen's hassocks and tabouret tables, and various items designed by Garth and Ada Louise Huxtable (including a set of service-ware comprising in excess of one hundred pieces) – were also, by association, raised to the level of art. Art, design and nature combined, therefore, to present an elite, sumptuous, commercial interior environment.

The Four Seasons was an exceptional restaurant that brought late-modern commercial architecture, modern design, art and nature together in its interior for a select audience. Dallas's shopping mall, known as the NorthPark Center, which opened in 1965, employed similar strategies but aimed them at a larger section of the public. While the restaurant was promoting a new modern eating experience, the Dallas mall set out to create a new modern retail experience that combined shopping with leisure, pleasure and social activity.

The task of designing NorthPark's flagship retail outlet, the Nieman Marcus store, was originally granted to Eero Saarinen. When Saarinen died in 1961, however, the work was taken over by Kevin Roche, who was working in Saarinen's office at the time and had been collaborating with him on the design. Roche had begun his studies under Mies at the Illinois

Fig. 79 Water feature with fountain and planting outside the Nieman Marcus store in NorthPark Center, Dallas, designed by Lawrence Halprin, completed 1965. Photograph by the author, 2007

Institute of Technology in the late 1940s, but did not remain there long because he found the teaching lacked social engagement. He subsequently took a position in Saarinen's office in 1950.[15] The generational shift from Mies to Saarinen brought a transformation of inter-war modernism into a softer version that was infused with Scandinavian humanism. Roche felt more comfortable in the Saarinen office than he had under the tutelage of Mies. His contribution to the Dallas mall represented, therefore, a new phase of late-modern architecture that was comfortable with embracing nature in order to soften its hard materials and humanise its forms.

Urban and suburban malls led the way by facing inwards, and bringing the outside indoors.[16] The first enclosed mall – Southdale, located in a suburb of Minneapolis – had been developed in 1956 by a Viennese modernist émigré, Victor Gruen.[17] The aim had been to protect shoppers from harsh weather and to separate them from the crime and troubles of the world outside. Gruen's vision had been the creation of a modern agora, a heart of the community. Nature, in the form of potted plants and fountains, was included as a reminder of the more pleasant aspects of the world outside that had been left behind. The introduction of natural elements also helped soften the hard structure of the malls, created a homely atmosphere, added beauty (often reinforced by the addition of pieces of sculpture and works of art) and compensated for the otherwise overtly commercial function of the spaces in question. Southdale's mall boasted an aviary full of songbirds and a fishpond, as well as a great deal of foliage. A skylight was included to allow sunshine to penetrate, reassuring visitors of the continued existence of an outside world. Although malls were public spaces, the surveillance systems that were installed in them facilitated a high level of control and ensured that unwanted visitors could not enter.

The Dallas mall was the brainchild of the art collectors and property developers Ray and Patsy Nasher, and is located on the northwest corner of Northwest Highway and Central Expressway.[18] Built on a site of 97 acres, it was completed in August 1965 and contained eighty new retail stores. It was the largest climate-controlled retail centre in the world in its day and it went on to win the Association of Interior Architects' Design of the Decade award. As with the Four Seasons, creating the Dallas mall required input from a group of experts. The architects, Harwood K. Smith and E. G. Hamilton, were a local team.[19] They worked alongside Kevin Roche, the planner Marvin Springer, and the landscape architect Lawrence Halprin.[20] In late 1962 the whole team was convened by Ray Nasher for a five-day brainstorming session. The scheme that emerged was an L-shaped configuration with an anchor store at the corner and each end. Varied sequences of interior promenades and courts connected them. It was Hamilton's decision to call NorthPark a 'center' rather than a 'mall', to give it a different identity, and he wanted the interior to be conceived as a series of spaces, as in an art gallery, rather than a corridor. The design was strongly unified, and included abundant natural light. Individual stores were subjected to rigorous design review to ensure that cohesion was maintained. Hamilton, who was a great admirer of Mies, was keen that the centre be perceived as 'architecture' rather than a 'stage set'. He came up with a simple palette to draw everything together, which consisted of cream glazed bricks, cast stone and concrete. The result was a soft, contemporary building that was neither monumental nor severe.

Just as Linn had been involved with the Four Seasons, so a landscape architect was needed to assist with the interior at NorthPark. Lawrence Halprin had opened his San Francisco practice in 1949. He had a particular interest in enlivening cities through the addition of landscapes. Together with the local landscaper Richard B. Myrick, he undertook the design of the interior courtyards, the landscaping and the planters at NorthPark. Patsy Nasher was also interested in the landscaping and planting, and worked closely beside Halprin. Halprin was especially fond of using water, and in Dallas he worked on the pond project outside Nieman Marcus and created two more fountains outside the Titche Goettinger and J. C. Penney stores (fig. 79). The first comprises a Zen-like fountain, a footbridge and four large planters made of dark glazed bricks on which children love to slide.

Fig. 80 Joel Shapiro, *20 Elements*, 2004–5. Wood with casein, 121.6 × 132 × 85 in. NorthPark Center, Dallas. Photograph by David Teagle, 2006

Fig. 81 Plants in pots at NorthPark Center, Dallas. The shallow pots allow the planting to be viewed from above. Photograph by the author, 2007

The fountain outside Titche Goettinger is octagonal and features a water-and-light show, while the last is rectangular and boasts round terracotta planters filled with chrysanthemums around its edge. The fountains provide an animated counterpoint to the relative calm of the retail concourses. Halprin's work also played an important part in the creation of social spaces at NorthPark in which people could congregate.

From the outset the large central area of the mall was filled with numerous potted plants. Anthuriums, bromeliads, grasses and mother-in-law's tongues were among the many varieties that were included. They were arranged in a highly stylised, modern manner and sat alongside contemporary art from the Nashers' collection, which included pieces by Barry Flanagan, Antony Gormley, Henry Moore, James Rosenquist, Joel Shapiro, Frank Stella and Andy Warhol, among others (fig. 80). The use of polished concrete on the mall floor added to the dramatic visual effect of the plants, which were embedded in a variety of circular, square and rectangular planters, arranged in regular groupings.[21] Most of the planters were wide and low so that the flowers and plants they contained showed to good effect from a higher level in the mall (fig. 81). Much attention was given to the geometry of the clusters of potted plants, which provided large blocks of colour and texture. There was a strong artistic and architectural flavour to the arrangements. In addition to those positioned in the centre of the pedestrianised areas, plants were also used throughout the mall to function as space dividers, barriers and screens. As at the Four Seasons, they were changed four times annually according to the time of year.

The Ford Foundation Building at 320 East 43rd Street in New York, one of the most polluted areas of Manhattan, opened in 1968 featuring an indoor garden in an enclosed atrium. Although the building was a site of office work rather than of consumption or leisure, it also sought to draw employees during non-working hours, and members of the public. It adopted the idea, inspired by the shopping mall, of turning itself away from the street and of creating its own outside world within a newly built structure. In the words of David Gissen, it built on the nineteenth-century idea of using 'indoor greenery as a counter to urban industrialization in Western metropolises ... offering a parallel indoor world ... that contrasted with the environmental depravity of the industrializing city'.[22] However, as Gissen also points out, the inclusion of inside greenery additionally increased the value of the space in question.[23]

The Ford Foundation Building consists of twelve storeys of glass, rusted steel and pink granite and contains the oldest large interior atrium garden in the United States.[24] Unlike the interior of NorthPark, however, the glazed wall of the atrium partly interfaces with the street. It presents itself as a public space, as well as a facility for the workers in the Foundation's offices. More obviously than the mall, it can also be seen as a giant greenhouse, or winter garden, filled with plants (figs 82 and 83). Given that the building was designed as a cube with an atrium on its southeast side, its essential function

Fig. 82 Looking down on the indoor garden at the Ford Foundation Building, New York, designed by Kevin Roche and John Dinkeloo, completed 1968. Photograph by Dick Lewis, 1979

as an office was separated from the garden. Senior employees worked in enclosed glass cubicles that looked out onto the interior garden, enabling them to see each other. There was a strong utopian aspect to the garden as it was meant not only to unite the staff but also to provide a peaceful area in which they could relax when they were not working (though there were no benches, and eating there was prohibited). It was also intended to create a green oasis in a city that was in poor economic and environmental condition, and where the trees were dying as a result of pollution.

Kevin Roche and John Dinkeloo, who had already worked on the Nieman Marcus store in the NorthPark Center, were the architects of the Ford Foundation Building. Roche had a strong interest in public spaces and public gardens. He also saw the limitations of pre-war modernism. The decision to include a garden in the complex reflects his desire to link architecture with nature. However, it was a tamed, controlled version of nature that interested him, one made possible by technology. Also, as at the Four Seasons and NorthPark, Roche understood that nature could be used to add seasonal variability into otherwise static buildings.

Landscape architect Dan Kiley was introduced into the Ford Foundation indoor-garden team (with whom Roche had already worked at the Oakland

Fig. 83 A fiddle leaf fig *(centre)* and other planting in the indoor garden at the Ford Foundation Building. Photograph by the author, 2007

Fig. 84 A weeping fig *(centre)* and other planting beside the steps in the indoor garden at the Ford Foundation Building. Photograph by the author, 2007

Museum, California).[25] Kiley had been apprenticed to the landscape architect Warren Manning in the 1930s, and had gone on to study at Harvard alongside Garrett Eckbo. After the war, in 1955, he had worked on the garden of the Miller House in Indianapolis, which featured interior work by Eero Saarinen and Alexander Girard.[26] A further continuity was established when the horticulturist and interior landscaper Everett Conklin, who had been involved in the Four Seasons project, was brought in to advise on, install and maintain the greenery in the Ford Foundation Building's indoor garden. The team worked together to create a piece of late-modern commercial architecture that did not merely include plants, but was largely defined by them.

The skylit, curtain-walled garden that the group created is located in a space 160 feet tall (equal to ten storeys). Its staircase (fig. 84), which is broken into three sections and from which terraces flow, leads down to a square pool of water (figs 85 and 86). There is a 13-foot height difference between the entrances on 42nd and 43rd street. Planters were added on the third, fourth and fifth floors of the building. Following Kiley's vision for the space, and with the help of Conklin, nearly forty trees, 1,000 shrubs and over 22,000 vines and ground-cover plants were installed in the atrium.

Fig. 85 The pool in the indoor garden at the Ford Foundation Building, New York, 1979. Photograph by Dick Lewis

Fig. 86 An umbrella tree stands in the pool in the indoor garden at the Ford Foundation Building. Photograph by the author, 2007

The aim, in the early years, was to use a range of temperate plants. Eight magnolia trees, each 25–30 feet tall, were brought from Richmond, Virginia, to form the core of the garden. Jacaranda trees, an evergreen pear, a red ironbank and a Japanese cryptomeria were also planted there, as were several camellias and six varieties of azalea, among other plants and flowers. Ornamental ferns and grasses were also used extensively. The lighting, irrigation and drainage were carefully calculated following ideas developed by the Dutch biologist Frits Went. Ada Louise Huxtable described the garden as 'a horticultural spectacular and probably one of the most romantic environments ever devised by corporate man'.[27]

Kiley deliberately over-planted the garden, creating a jungle in which plants would survive as best they could. However, the high humidity and low light levels (the garden is hemmed in by neighbouring skyscrapers, so receives little sun) were not right for the first plants that were chosen and they eventually had to be replaced with subtropicals that were tougher and had a better chance of survival. Ten years after the garden was completed Conklin explained how they had tried, and largely failed, to grow trees of the southern temperate zone in a tropical or semi-tropical environment.[28] At the time, six of the magnolias were still alive, as were all the camellias, but the three Japanese cryptomerias had died.

Conklin's article was much more than a statement about the condition of the Ford Foundation garden, however. He also discussed the future of indoor planting, explaining that:

Natural Late Modernism

Fig. 87 Towering bamboo in the courtyard garden of the IBM Building, New York, designed by Edward Larrabee Barnes & Associates, opened 1983. Photograph by the author, 2007

> the interior planting specialist of today works very closely with the interior designer, architect, and landscape architect and has not only the ability to interpret the designer's every wish into practical plantings, artistically and functionally arranged, but to maintain those plantings on an unconditional plant replacement guarantee basis, thus eliminating all worries and the responsibility on the part of the owner.

He went on to say that he did see a future for the interior planting business, though he was not sure whether it would be in the hands of the outdoor planting people and nurserymen or the interior planting specialists. He sensed it would be the latter. In either case, 'Indoor plants and trees are here to stay, especially in our city buildings', he proclaimed prophetically.[29]

As well as predicting what was soon to become a massive industry, Conklin was far-sighted enough to understand that a huge shift was taking place in terms of the relationship between human beings and nature. He went as far as to claim, again presciently, that 'In the absence of plants and flowers in some of our interiors we may be producing undesirable side effects that, if allowed to proceed unchecked, may result in the erosion of the psychological environment, the erosion of human life.'[30]

Through the 1960s and beyond, countless large public and semi-public buildings across urban America had plants and flowers introduced into their inside spaces. The IBM Building, opened in 1983 at 590 Madison Avenue in New York, was among the many to contain numerous plants – huge bamboo trees, in this case – in their indoor atria (fig. 87). 'By the late 1980s', wrote David Gissen, 'the sheer volume of plants being grown throughout New York City and other cities made the indoor atmosphere of late-modern architecture a new lens through which to re-evaluate the natural and productive capacities of architecture more generally'.[31] As ever, nature inside was used to a number of different ends, among them softening and humanising the monumentality of late-modern architecture and encouraging consumption by making it a more relaxing and enjoyable experience. Whether these were 'corporate arcadias' for the many, or whether they were there only for the chosen few and, once again in Gissen's words, simply 'reinforce[d] the trade systems and political relationships that underpin[ned] the modern experience' was widely debated.[32]

Like the NorthPark Center and the Ford Foundation Building, many other buildings of the era turned their backs on the urban problems outside their walls and used nature inside as a means of creating an enclosed and private fantasy world. One such was the Hyatt Regency Hotel, the subject of the following chapter, which was designed by John Portman and opened in Atlanta in 1967.

CHAPTER 8

The Living Room in the City: The Hyatt Regency Hotel, Atlanta

John Portman ... usually designs the space as a courtyard but furnishes it as a living-room.[1]
Michael J. Bednar

By the late 1960s, the large plant-filled atrium building had come into its own, thanks largely to the work of the Atlanta-based architect John Portman, who was known for the design and construction of urban atrium hotels and commercial megastructures that seamlessly linked offices and retail spaces with public environments. His work in that field provided a model for many of the late-modern, urban, public and semi-public interior spaces with atria – from hotels to office blocks to shopping malls – that, festooned with potted plants and flowers, became a familiar sight across the globe in the late twentieth and early twenty-first centuries.

The way in which Portman brought nature, in the form of plants, flowers, trees, water and, occasionally, live birds, into the interior built on the earlier role of nature within nineteenth-century domesticity. If the city was a metaphorical house, he explained, then his vast hotel atria, like nineteenth-century parlours before them, were spaces in which refuge, repose and pleasure could be found and the senses indulged. At the same time those same atria could be understood as city squares and microcosms of the city itself. Portman linked his buildings together with high-level covered walkways that enabled people, as they had in the nineteenth-century public leisure centres, to move from place to place unaffected by the weather. Unlike in southern California, in Atlanta the weather was more an enemy than a friend.

Portman graduated from the Georgia Institute of Technology in Atlanta in 1950 and formed his architectural practice in 1953, joining H. Griffith Edwards three years later to form Edwards & Portman Architects.[2] As a final-year student he was taught by Frank Lloyd Wright, when the revered

Fig. 88 The front cover of John Portman and Jonathan Barnett, *The Architect as Developer* (New York: McGraw Hill, 1976)

elderly architect was a visiting tutor at Atlanta. The experience was a formative one for Portman, who subsequently drew heavily upon Wright's notion of 'organic architecture', the idea that harmony between humans and nature can be achieved through an approach that is sympathetic to both, blending buildings with their surroundings to create a harmonious whole (see Chapter 4). Portman was to take the organic idea forward into a popular and overtly commercial context, transforming it significantly in the process. In his 1976 book, *The Architect as Developer*, which he co-authored with Jonathan Barnett, Portman explained that he was also influenced by the work of Louis Kahn and Eero Saarinen (fig. 88). He clearly saw himself working within the tradition of post-war American late modernism, which sought links between architecture and the natural world and endeavoured to make buildings attractive to people by appealing to their senses.[3]

In 1961 Portman designed the Merchandise Mart in Atlanta, his first large project, which led him to transform a vast area of land in downtown Atlanta over the next half century. He approached that gargantuan task as both a real-estate developer and an architect, a dual role that he sustained throughout his career. The building, which initiated his hugely successful plan to bring trade and finance into the city, was topped by a restaurant containing numerous plants.

The two homes Portman built for himself and his family – Entelechy I and Entelechy II – allowed him to try out, in a private context, many of the architectural themes and strategies that were to characterise his public work over the next decades. In those projects the architect demonstrated his determination to use trees and plants, and, importantly, water, in the inside spaces of buildings. The strategies he used to blur the boundaries between inside and outside spaces, between the public and the private spheres, and between domesticity and its opposite, were also realised in his own homes. The introduction of nature inside was key to the implementation of those strategies.

Built in 1964, Entelechy I (its name means 'realising one's potential') was a two-storey building constructed around twenty-four skylight-capped hollow columns. Its spatial composition relied upon the inclusion of trees, plants and water. The living areas occupied the spaces between the columns, while the columns themselves contained utilitarian features, including staircases, a powder room and cupboards. A few contained potted plants. Portman set out to create a close link between the building's exterior and its interior through the manipulation of natural elements. Trees were planted in curved bays that were exposed to the sky and separated from the interior proper by floor-to-ceiling windows. At ground level Portman brought water right inside his building in the form of a shallow pool that ran from its north side through to its south side, between the fourth and fifth columns. The pool served to separate the formal living areas from the family's private spaces, flowing as it did directly into the little bays that jutted out between two columns on both sides of the building. Within the pool Portman created a number of islands. The central one was circular and supported a dining table and chairs, while the island on the south side of the house featured three large trees used to reinforce the idea that the outside had found its way inside.

Larger trees were used as landscaping around the swimming pool located at the rear of the house on the south side, serving, once again, to establish continuity between the exterior and the interior. Portman's creation of a dialectic between the built structure and nature inside contrasted soft with hard, green with neutral tones, dynamic (as the light played on the leaves) with static, and the ephemeral with the permanent. It created an animated conversation and gave a sensuous and emotional edge to his use of an otherwise conventional late-modern architectural vocabulary dominated by the concepts of form, space and light.

In the 1980s Portman went on to build Entelechy II, a seaside home for himself and his family on Sea Island off the coast of Georgia. The house consisted of four linked pavilions, supported, once again, by hollow columns. Having learnt much from inter-war modernism, Portman introduced terraces, balconies, roof decks, skylights, open courtyards and an indoor pool; together, they established considerable ambiguity between the inside and the outside of the building. The countless trees and plants that filled the building and covered the surrounding land rendered that ambiguity even more complex. Describing the house, the Italian critic Aldo Castellano explained that 'space was the goal'. Inasmuch as it felt as if a centrifugal force was operating inside the building, directing its spatial composition from the inside to the outside, he likened its composition to a 'controlled explosion'.[4] The result was a fragmentation of space that required a significant level of architectural control.[5] The strategies that Portman developed to exert that control included the introduction of a dialectic between nature and culture; between, in his own words, 'vegetation and construction'.[6]

In Entelechy II Portman used nature to create a positive tension that gave his spatial composition coherence. Equally importantly, however, his use of nature confirmed that his version of late modernism embraced the world of the senses. To emphasise that, he refused to trim the vegetation back and, as it flourished in the near-tropical setting of the house, ultimately engulfing the architectural construction and rendering it almost invisible, it gave the building an increasingly wild appearance. Inside the house, green potted plants were used to break up the geometry of its internal spaces; trees hovered over the first-floor decks that overlooked the swimming pool at the rear of the house; and in the master-suite living room a rampant ivy flowed down the front of a concrete mantelpiece, rather like the one in the painting by Ludwig August Smith (see fig. 16) and in the Aaltos' home in Helsinki (see fig. 47). Castellano explained that nature in Entelechy II was never out of control, however, and that culture, in the form of architecture, always had the upper hand:

> The green is ever present. It creeps down the walls, it climbs up and down from large and small flower pots, it appears here and there, as the bushes in a heath, but it never covers, nor could it, the immediate perception of the free articulation of architectural space.[7]

As the vegetation grew, so the interior of the house increasingly recalled the Victorian ideal of the domestic jungle, evoking a vision of its inhabitants having to fight their way through undergrowth to be able to sit on a sofa in their parlour.

Back in 1965, Portman's wish to bring nature and light into his public spaces as well as his private ones had been apparent in his design for the Greenbriar Shopping Center in Atlanta, which was described at the time as 'an extensively landscaped interior mall and courtyard system'.[8] Greenbriar was the city's seventh mall, but only its third fully enclosed shopping complex. At its centre it featured a sunlit mall-way containing large concrete statues of animals, two fountains and an aviary. The large palms planted in enormous white plastic tubs that ornamented the open spaces of this consumers' paradise appeared fresh and appealing, and provided a sense of the outside even as the mall gave shoppers security from exterior threats and inclement weather – especially the humid Georgian summers. Early signs of a desire to introduce sky-lit interior courtyards into his buildings were apparent in Portman's design of the same year for a low-cost residence for

Fig. 89 The atrium lobby at the Brown Palace Hotel, Denver, designed by
Frank E. Edbrooke, opened 1892; photographed 1987. Photographer unknown

the elderly in Atlanta, the Antoine Graves Home, which had two such interior spaces. The public was not invited in to enjoy those light interiors, however.

The atrium – a term originally used to describe the central open space in a Roman villa – had nineteenth-century precedents in the United States, notably in the fashionable, Renaissance-style Brown Palace Hotel in Denver, which opened in 1892 (fig. 89). Described as an 'atrium lobby', the first space visitors encountered was an eight-storey void surrounded by cast-iron balconies decorated with ornate grillwork panels. Its architect, Frank E. Edbrooke, had created a dramatic indoor environment; among many other luxurious details, it featured onyx imported from Mexico, which undoubtedly impressed travellers to Denver.[9] The palms in the Brown Palace atrium played a similar role to the potted plants in a nineteenth-century domestic interior, adding a sense of the exotic while also making guests feel at home.

The Living Room in the City

Many other early twentieth-century American hotel lobbies – such as the one in Memphis's 1925 Peabody Hotel, which has been described as the 'living room of Memphis', and the example in New York's Plaza Hotel (see Chapter 3) – were similarly impressive and luxurious. There were also precedents for the exposed lifts that Portman used in several of his buildings. They included the open-cage elevator on the Eiffel Tower built in 1889, the lift that was installed in the El Cortez Hotel in San Diego in 1956 and the example on the Space Needle created for the Seattle World's Fair in 1962.

In his historical account of the atrium, Michael J. Bednar has pointed out that, through the ages, it has existed in both private and public spaces, performing several roles ranging from parlour to winter garden.[10] Portman's hotel atria were an interesting mix. Theoretically, they were public spaces that anyone could enter, but they were only really accessible to those who could afford to stay in the hotels, eat in their restaurants and drink in their bars. Metaphorically, however, they were private, domestic spaces in which individuals could find safe and pleasant havens from the threatening streets outside. There were also precedents for buildings, including hotels, with dramatic and dynamic entrances and means of movement through them. Portman himself cited the examples of Morris Lapidus's dramatic curving lobby stairs at the Hotel Fontainebleau in Miami, opened in 1954, and Frank Lloyd Wright's use of spiralling ramps to create open space and movement in his Guggenheim Museum in New York (1959).[11]

The story of the Hyatt Regency Hotel, Atlanta (or, as it was originally known, 'the Regency Hyatt House'), went back to the early 1960s when, given the success of the Merchandise Mart, and while working with the Phoenix Investment Company and Trammell Crow, a Dallas investor, Portman decided that Atlanta needed a good hotel. Ground was broken in 1964, but halfway through the construction process the project was sold to the Hyatt Corporation. Under the leadership of Jay Pritzker, its chairman, and Don Pritzker, its president, the Hyatt Corporation bought the Atlanta hotel in April 1966 for 16.5 million dollars. Strikes delayed the opening of the new hotel until May 1967. By the time it was completed it had cost 18 million dollars.

The Hyatt Regency was an enormously ambitious project that set out to redefine Atlanta. The last hotel built in that city had been the Jefferson in 1921. By the mid-1960s the concerted effort of Atlanta officials, including the mayor and the members of the Chamber of Commerce, and the contribution of powerful and influential men such as John Portman, had begun to transform the city. The scale of Portman's atrium at the hotel reflected the level of his ambition. It was a huge and complex construction – the courtyard contained 3 million cubic feet of space and was over 200 feet high – in which space, light, colour, texture, movement and sound were combined in a highly sophisticated manner.[12] Several commentators remarked that the Statue of Liberty would fit into it. Tamed nature (plants, flowers, water and live birds) blended with culture (the built structure, the furniture and the furnishings) to create a dynamic – Portman used the word 'kinetic' – space that stimulated the senses.[13]

The Hyatt Regency was to establish Portman as an architect of national significance and, soon afterwards, of international notoriety. Unlike the Antoine Graves Home, the Hyatt Regency consciously set out to encourage visitors – whether or not they were guests at the hotel – to come off the streets and enjoy the pleasures it had to offer, among them its twenty-two-storey open atrium filled with a variety of attractions including the exposed plexiglass lifts that travelled up and down a central hollow column. 'Because space is at a premium in the city', Portman explained, 'the atrium would offer refreshment to those who came from the street.'[14] He must have been aware that there were racial tensions at that time in Atlanta, and that, although the city was playing an important part in the growth of the civil rights movement in the 1960s, and was moving towards desegregation at a significant pace, the street was not always the easiest place to be.

That overt rejection of the urban exterior brought Portman a great deal of hostile criticism in the 1970s and 1980s. He was accused of spurning the spaces of the city and the life that went on in them, and of turning his architecture to look in on itself (like the NorthPark Center in Dallas and the Ford Foundation Building in New York, discussed in the previous chapter). Some critics also felt that his spaces were lifeless and inauthentic. That inward-looking quality made the hotel a kind of protected citadel, one which, unlike the Villa Tugendhat and its porous relationship with the land on which it sat, was defined by a break with the world outside. Yet at the Hyatt Regency the energy and life of the street were largely reproduced inside. As part of his ambition to create an animated and sensuous 'city inside', Portman continued to develop the dialectic between built form and nature that had, by that time, become an integral part of his approach to architecture. 'We wanted to awaken all the senses so there was music, fountains and birds. We wanted to integrate nature into the city', he explained.[15] In spite of the scale of the project, and the late-modern language of its architecture, he also set out to evoke the humanity – the animation, the intimacy, the domesticity, the cultural aura and the spiritual ambiance – of the Victorian parlour in which people and nature had cohabited in an interior space that united them, and in which their identities were formed. Recognising that, Portman explained: 'We wanted [it] to become the living room of the city'.[16]

The scale and the grandeur of Portman's lobby were unprecedented, and his atrium was widely heralded (however falsely) as the first of its kind. 'This is the largest hotel lobby in history ... perhaps the only space of its type built since the Renaissance', wrote an enthusiastic journalist in the 25 June 1967 edition of *The Atlanta Journal*.[17] Portman demonstrated his ongoing debt to Frank Lloyd Wright by emulating one of his tactics from the Johnson Wax Headquarters Building of 1936–9 (in Racine, Wisconsin). Just as Wright had made people enter his building through a low side entrance

Fig. 90 Looking up through the atrium of the Hyatt Regency Hotel, Atlanta, designed by John Portman, opened and photographed 1967. Photograph by Alfred Eisenstaedt

off a parking lot, a strategy which had cleverly emphasised the height and openness of the enormous workroom inside, so Portman forced his visitors to enter the Hyatt Regency through a stark, low, dimly lit tunnel that added to their sense of shock at their first sight of the vertical, light, air-conditioned atrium topped by forty-seven skylights (fig. 90).[18] The tunnel was, one commentator remarked, 'a deliberate preparation for the visual surprise that will await the guest when he emerges at the end'.[19] Visitors to the hotel in 1967 reported their gasps of surprise as they entered the atrium.[20] Some dubbed it a 'fabulosity' and others 'the eighth wonder of the world'.[21] Both *Life* and *Time* magazines ran celebratory features about the hotel when it opened, and the renowned architect Edward Durrell Stone flew into Atlanta especially to see it – and straight back out again afterwards.[22]

Visitors' senses were clearly bombarded by the huge space they encountered when they entered the atrium. The journalist writing in *The Atlanta Journal* in 1967 described that dramatic void as follows: 'On all four sides rise the vine-covered interior balconies of the rooms, while the elevators move up and down on their shafts in full view like

Fig. 91 Frank Lloyd Wright, Fallingwater (the Kaufmann Residence), Pennsylvania, completed 1935. Photograph by Allen Brown, 2017

gems on a necklace.'[23] Six Australian umbrella trees, each 30 feet high, and numerous other plants and flowers – including chrysanthemums and the climbing philodendrons that flowed from the balconies – combined to create visitors' first impressions as they exited the dark tunnel. Once again Portman was making a connection to the work of Frank Lloyd Wright, this time to Fallingwater, his 1935 house in Pennsylvania (fig. 91) designed for the Kaufmann family, the large cantilevered terraces of which had sumptuous vines growing over them. Even earlier links can be made to Greene & Greene's Gamble House (see Chapter 4), where vines flowed down the façade, linking the building to the garden below. Given that the built-in concrete planters were embedded in the structure of the building, the vegetation that Portman introduced into the Hyatt Regency must have been an integral component of his architectural vision, rather than a decorative afterthought.

Anticipating Aldo Castellano's later description of Entelechy II, in 1967 an Atlanta journalist described the Hyatt Regency atrium as 'a total explosion of space'. He added that the Hyatt Corporation had bought 'the idea of the grandeur of space'.[24] The Hyatt Regency was a new kind of hotel, one that

The Living Room in the City

Fig. 92 A seating area near the entrance to the atrium of the Hyatt Regency Hotel, Atlanta, with one of the glass-walled elevators in the foreground, 1967. Photograph by Alfred Eisenstaedt

rejected the traditional idea of the double-loaded corridor hotel and, instead, was constructed around an inner courtyard like a piazza or village square. This likeness was reinforced by the use of quarry tiles on the floor, which were roughly laid in a fanned pattern to resemble cobblestones. Portman dedicated much effort to the way in which space was articulated in the interior of his atrium, often defining it through the use of plants and flowers.

The details of the interior design of the public areas in the Hyatt Regency were entrusted to Roland Jutras, a well-known creator of hotel interiors at the time. His design business, Roland Wm. Jutras Associates Inc., was based in Boston. His intention was to create a sense of excitement and dynamic activity, and to provide a variety of eating, drinking and relaxing spaces for guests and other people who may have walked in from the street. On entering at street level, atrium visitors were first confronted by two of the large Australian umbrella trees, which framed their entrance, and, in front of the trees, a large relaxation-cum-conversation space comprising four long, low sofas positioned on three sides of a square (fig. 92). When the hotel opened, black leather and chrome chairs designed by Marcel Breuer were positioned in the middle of the rectangle thus formed, facing the outer sofas, while two low glass-and-metal coffee tables stood between the chairs and the sofas. In the rear two corners lamps, set on small marble-topped tables gently illuminated the area, and the whole ensemble sat on a bright red rectangular rug. A rectangular black planter containing yellow chrysanthemums, subsequently to become a hallmark of Portman's interior spaces, stood between the two rear sofas. That strikingly horizontal, elegant and relaxed assemblage of furniture contrasted sharply with the dramatically vertical, 70-foot-high fountain which rose behind the conversation area from its root two floors below, taking the visitor's eyes up into the space of the atrium and to the vine-covered balconies lining its four sides. Water travelled back from the top of the fountain along transparent plastic strips, sparkling and bubbling as it went. More Australian umbrella trees flanked the fountain, again encouraging viewers to raise their eyes.

The 'Kobenhavn Kafe' (Copenhagen Café) was situated to the left of the conversation area, separated from it by a low partition along the top of which a stream gently bubbled (fig. 93). Copenhagen – one of Portman's favourite cities, which he had visited in the mid-1960s – strongly influenced the Hyatt Regency's interior. The café contained street lamps and figurines that had been handmade and painted in Copenhagen, as well as marble-topped and wrought-iron-framed tables and chairs similar to those Portman had seen in open-air cafés there. The architect later claimed that Copenhagen's Tivoli Gardens had been a revelation to him. He had seen people walking through Tivoli with smiles on their faces, a rare sight in Atlanta.[25] The 'city' he set out to create inside his hotel would, he hoped, provide an alternative kind of outside space that people would enjoy inhabiting. The fairy lights on the exposed glass lifts were also inspired by Tivoli, introduced to evoke a pleasure-garden experience for the hotel's visitors.

Fig. 93 The Copenhagen Café in the atrium of John Portman's Hyatt Regency Hotel, Atlanta, 1968. Photographer unknown

Behind the fountain and suspended by a cable from the ceiling of the atrium, Portman and Jutras introduced an umbrella-like form over a bar and lounge area. Made of wrought iron cut into a quasi-Art Nouveau pattern, with plexiglass infilling, the form evoked Frank Lloyd Wright's stained glass and injected a sense of nostalgia into the interior. One writer called it 'a giant Tiffany lamp', while another referred to the metal tracery used by Louis Sullivan on his buildings.[26] The cocktail bar sheltered by that giant cover was aptly named the Parasol Lounge.

Much of the ground floor on the right-hand side of the lobby was taken up by the five metal and plexiglass lifts, supported by hollow columns, which provided one of the most dramatic features of the atrium. Supplied by the Otis Elevator Company, with the capacity to carry twenty people each, they allowed their passengers to experience the atrium's full grandeur and drama (fig. 94). 'It's the best ride since Coney Island', one visitor exclaimed in 1967.[27] The lifts travelled at 700 feet a minute. Two were enclosed, to accommodate fearful passengers. They went all the way up to what was perhaps the hotel's most impressive feature: its revolving, circular rooftop

Fig. 94 Looking down on the atrium of the Hyatt Regency Hotel, Atlanta, with glass-walled elevators in motion on the columns, 1966. Photograph by G. E. Kidder Smith

restaurant, the Polaris, which, illuminated in blue at night, resembled an alien vessel that had landed from outer space. While one writer linked the atrium to the Tivoli Gardens and commented on 'the combination of nature and fantasy which is the special charm of that Edwardian pleasure garden', another saw a more local influence at work on the lift cars, claiming that 'the lights of the elevator echoed the river boats from Currier and Ives and Southern mythology'.[28]

The public area in the lower level of the hotel – dubbed the Terrace Floor, as it led onto the outdoor pool area – provided visitors with yet more extravagant experiences, among them the Club Atlantis, the Phoenix Ballroom and Hugo's Gourmet Restaurant. The last featured a central lighting fixture composed of more than 2,000 glass test tubes, and offered, among other special dishes, whole lobster baked in its shell. The sense of glamour and luxury of the first floor was extended to the guests' and visitors' dining experiences. An aviary, which extended into the Parasol Lounge on the ground floor, and which was filled with exotic birds including toucans, parrots and macaws, had its base on that level.

Above all else, the public spaces in the Hyatt Regency assaulted the senses, especially those of sight and hearing. A writer in the July 1967 edition of *Interiors* described the hotel as 'a grand dream world of trees and twinkling lights and movement and people and sparkling fountains'.[29] The atrium performed that function primarily through the architect's dramatic articulation of space and form, and through his subtle use of colour, texture, movement and sound. The last was introduced into the atrium through background music, the bubbling of the fountains and the screeches of the exotic birds. In 1976 Portman explained that:

> I have at times put cages of birds into large spaces. In the Embarcadero hotel in San Francisco I have also introduced a 'sound sculpture' by Bernhard Leitner that creates the impression of a flock of birds flying in, alighting on the trees, and bursting into song. I use elements of nature to make a connection between the built environment and the human psyche.[30]

Water was very important to Portman. He had used it in a number of his projects before 1967, and he particularly liked the sense of movement it created, the sound that it made while moving, and the effects that were achieved when it was hit by light. Texture was equally important to him, and he provided it in a multitude of ways, including the rough surface of the concrete used throughout the Hyatt, and the leaves of the trees and plants that he added. The colour scheme of the lobby combined red, yellow and green as dramatic highlights to contrast with the neutral colours of the rough concrete surfaces and the quarry-tiled floor. The red came from the rugs that demarcated particular areas, including the static settings for conversations, which were thus differentiated from the circulation areas that

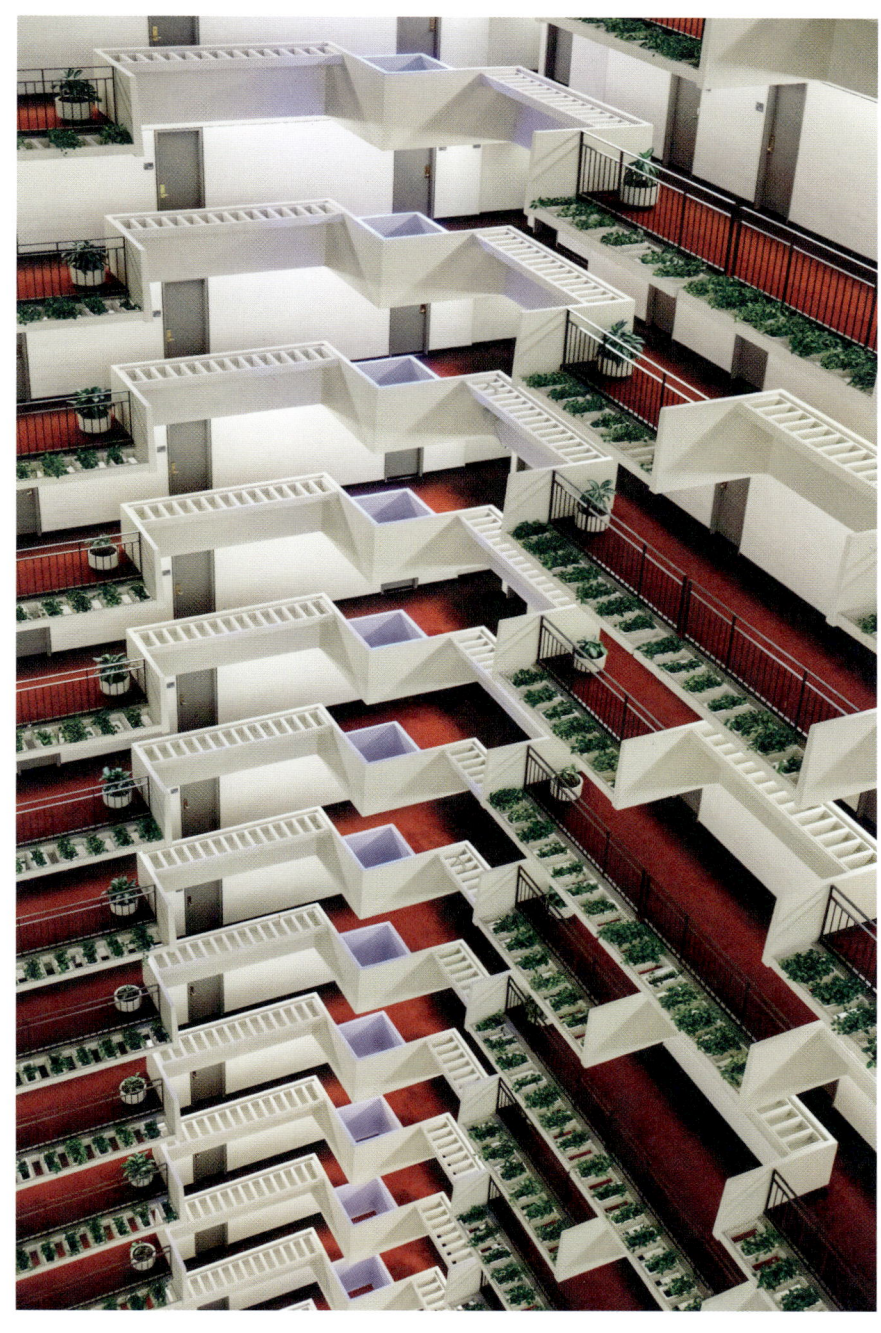

Fig. 95 Philodendrons in integral planters ornament the balconies that overlook the atrium of the Hyatt Regency Hotel, Atlanta, 2017. Photograph by Nick Higham

had no rugs on them; the yellow came from the potted chrysanthemums that appeared in a number of locations, including to the left of the Parasol bar, where seven pots were grouped together; and the green was provided by the upholstery of the café chairs and by the numerous plants that inhabited the space.

Plants and flowers played a key and integral role in the design and effectiveness of the Hyatt Regency Hotel's atrium. The writer for the 25 July 1967 edition of *The Atlantic Journal* explained that:

> a full-time gardening staff is just one of the unusual aspects of a spectacular concept in inn-keeping . . . a full-time gardener is in charge of caring for the larger tropical plants, and, at last report, he had a staff of two responsible for watering all the others once a week. It takes a full week to make the circuit.[31]

The article went on to say that, at the time of opening, the hotel contained around 3,000 plants, and that by the time the article had gone to press, that number was nearer 5,000. The total cost of the plants and their containers was approximately 100,000 dollars. Deciding where to place those plants was the task of the interior designer, Aubrey Parrott, who was clear about their role in the building. 'The plants were very necessary to humanize the scale of the building', he explained, 'to make the whole effect more comfortable and inviting. They provide a softening element in contrast to the bold and rather over-powering design. Without plants and trees, the architecture would appear somewhat stark and regimented.'[32] Through the variation in their size and scale the plants were also crucial in reinforcing the ambiguous metaphorical significance of the atrium space, the large trees suggesting that it was a city square and the smaller pots of chrysanthemums denoting its concurrent role as a living room or parlour.

The 2,400 vines used to line the balconies were supplied by Tom Mitchell of Tropical Gardens in Hapeville, Georgia. The species was the heart-leaf philodendron. An evergreen perennial with shiny rubbery leaves, native to tropical America, it is extremely hardy and can survive without much light. It was therefore ideally suited to the atrium of the Hyatt Regency. Individual plants in nine-inch pots were placed in rows in the grillwork planters built into the front sections of the concrete balconies that cantilevered out from the second to the twenty-second floors of the atrium (two on each side, giving eight in total on each level; fig. 95). The balconies were regularly staggered through the twenty floors to provide a visual progression of plants. A commentator writing in July 1967 explained that, 'When the hotel opened there were a hundred and ten Philodendrons to each floor, thirty to each large bay and ten to the smaller ones. Now all the bays are being filled.'[33] It had been the idea all along to use vines in this way, and the modular concrete structures used for

Fig. 96 Period postcard showing the planting at John Portman's Hyatt Regency Hotel, Atlanta, soon after its opening in 1967

the balconies outside the guest rooms were specially designed to accommodate them. The sheer quantity of vines flowing from the balconies provided a cascade of green vegetation, like water, down the inner walls (fig. 96). As well as adding colour, they provided texture and a strong sense of movement that contrasted with the static concrete structure of the atrium's frame.

Most of the other plants and trees used in the Hyatt Regency, including bamboo palms and striped dracaenas, came from Miami, as that was the nearest place where such a large number of plants was available at that time of year. In addition to the two Australian umbrella trees near the main entrance, and the two that flanked the fountain, another was positioned at the top-right corner of a sitting area to the right of the entrance tunnel. A final tree was placed in the far-right quarter of the lobby space behind the lifts and adjacent to a small bar and yet another sitting area. The strategic positioning of those six very large trees was crucial to the overall spatial configuration of the atrium and its definition as an urban space. Their round concrete planters measured five feet across, and each tree and planter weighed three tons.

Fig. 97 Indoor plants and trees in the office of Portman Architects, Atlanta, opened 1980s. Photograph by the author, 2011

As well as being key compositional elements, the plants were also included to add texture and colour and, as light fell on their leaves, to introduce a sense of dynamism. They contributed drama and human scale to the atrium interior and combined with the sound of the water and the birds to create a multisensory environment that appealed to all five senses simultaneously. It was a formula that Portman perfected at the Hyatt Regency, and that he employed in many of his projects over subsequent decades (fig. 97). It also provided a model for many other large atria constructed worldwide over the next half century, from hotels to office blocks to shopping malls.

When Portman and other architects working on large public buildings in the 1960s introduced plants and flowers inside, they knew instinctively that by adding a level of domesticity, softening the lines, creating immersive environments, playing with definitions of inside and outside, and creating indoor spaces of leisure and pleasure, they would transform their interiors and make them work as spaces of relaxation and entertainment for those who inhabited them. By the 1970s, however, financial investments in indoor planting schemes were so large that developers and architects were no longer willing to rely on their instincts alone. In order to provide hard evidence of the benefits of nature inside, several environmental and clinical psychologists, among a range of other people in the design-related industries, set out to prove that, as the authors of Victorian advice books had known a century earlier, indoor plants promoted both physical and mental well-being and detoxified indoor environments. From that point onwards, as interior landscaping became big business, the benefits of nature inside came to be appreciated not only for their own sake but also, increasingly, for their huge financial potential.

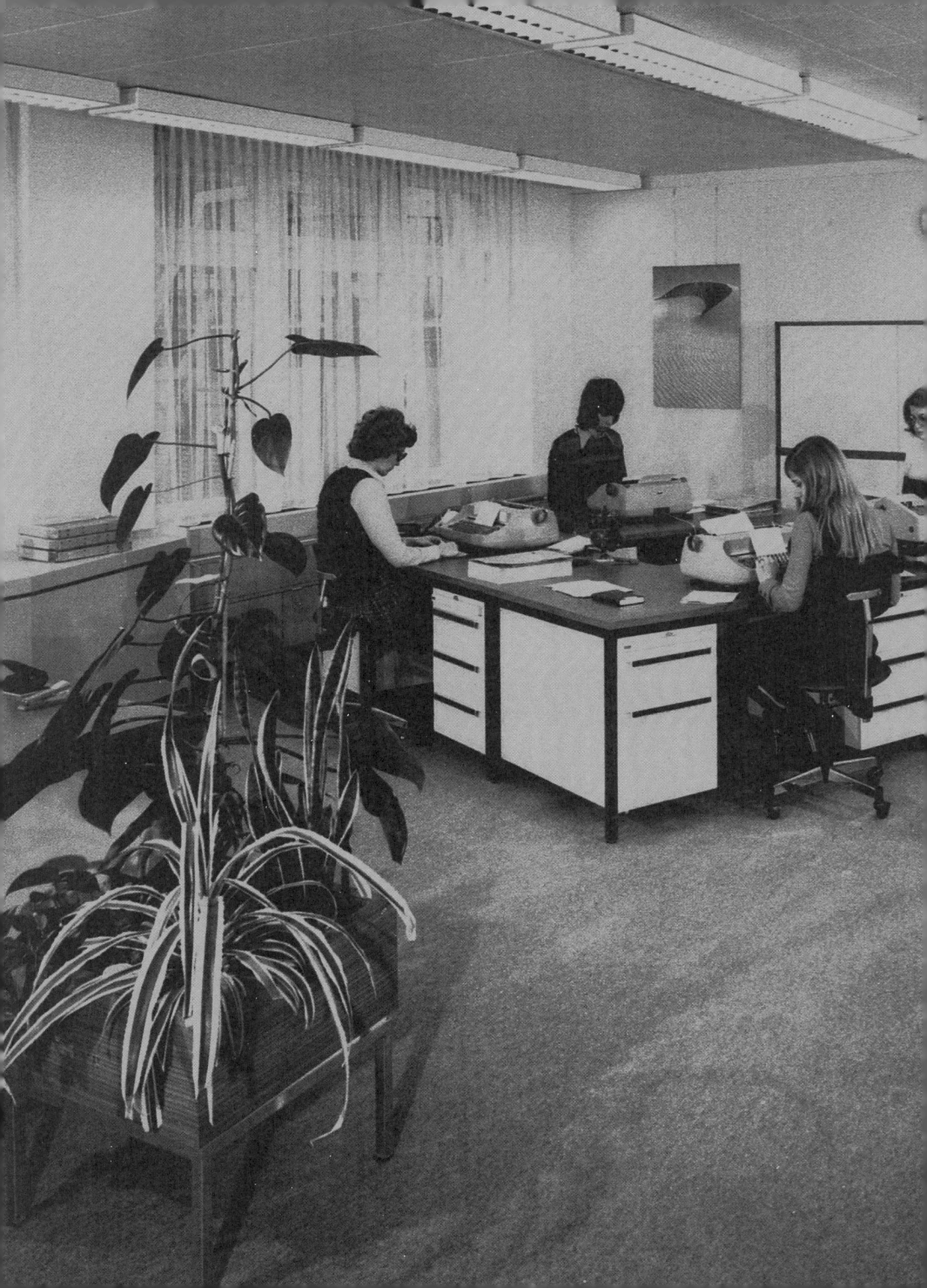

CHAPTER 9

The Benefits of Nature Inside

Nature in the West is instrumentalised as a mere means to human ends.[1]
Val Plumwood

As the previous two chapters have shown, the 1960s witnessed the introduction of nature into a range of large, public-sphere, indoor commercial environments in North America. Through the following decade that trend spread globally. In response to the financial investment made by the managers of those spaces, and in recognition of a growing general interest in the benefits of indoor plants, researchers in a range of adjacent fields set out to provide evidence of the effects of nature inside. The focus was on its potential to improve people's mental and physical health and well-being, as well as on its benefits for the environment. The commercial world's interest in the work being undertaken lay in the effect that indoor plants had on production in offices, consumption in retail and other commercial settings, and efficiency in hospitals, prisons and several other institutions. If people were happier and healthier, it was believed, they would work harder and better, spend more, and require less support, all of which boosted the financial bottom line.

    The desire to develop scientific models with which to examine the natural world recalls the rise of botanical science in the eighteenth century, and the work undertaken at that time by Carl Linnaeus and others to categorise plants systematically. By the 1970s, however, it was the psychological rather than the biological sciences that provided the baseline for debates and discussions about the benefits of tamed nature. In the eighteenth century, human beings had exploited tamed plants commercially by integrating them into the colonial economy. In the late twentieth and early twenty-first centuries, turning to science to provide a rationale for institutions and commercial operations to invest in indoor

Fig. 98 The front cover of A. B. Graf, *Exotic House Plants*, 10th ed. (East Rutherford, NJ: Roehrs, 1976)

plants represented a new form of exploitation of the natural world for economic ends. Of course, the growing number of nurseries that supplied indoor plants and flowers, and the indoor landscapers who installed and maintained them, also benefited from the expansion of their businesses.

On the back of this growth emerged a new profession, which was increasingly referred to as 'interiorscaping' or 'plantscaping'. While its practitioners were initially self-taught, formal educational systems were soon established. In addition to the members of architectural and landscape firms who worked on indoor foliage, several other individuals began to thrive as plantscapers. Everett Conklin was among the early pioneers, as was the New Jersey-based Roehrs family. Having started out in orchid-growing, the family had formed a business in 1869 and, by the early twentieth century, was running one of the largest nurseries supplying plants for indoor purposes. In the 1960s Julius Roehrs went to work in the Kentia palm industry in California. The Roehrs company also published books by the tropical-plant authority Alfred Graf (fig. 98), one of whose many claims was that the practice of bringing plants into the workplace had begun when women entered it (fig. 99).[2]

In 1974 an American landscape architect named Nelson Hammer convinced The Architects Collaborative, a company for which he was working at the time, to employ Conklin to help them install a lobby garden

Fig. 99 Natalia Danesi Murray (left), director of the American branch of the Arnoldo Mondadori publishing house, in her New York office with her secretary, 1955

in a hotel in the Middle East.³ While working for Dan Kiley a few years earlier, Hammer had become familiar with Conklin's interior landscaping at the Ford Foundation Building. He learnt a great deal from working alongside Conklin on the Middle East project, and afterwards went on to undertake a number of interior-landscaping projects himself, claiming that his main learning method was 'trial and error'.⁴ By 1992, though, he was able to observe that there were 'several excellent books and two national magazines devoted to the subject', as well as a number of university courses. Both developments suggest that plantscaping was beginning to acquire a serious professional profile.⁵

Richard L. Gaines's *Interior Plantscaping: Building Design for Interior Foliage Plants* (1977) was one of the earliest books to lay out the principles of the new professional practice.⁶ As the author explained, the terms 'interiorscaping', 'interior landscaping' and 'interior plantscaping' were used interchangeably at that time, though he personally favoured the last.⁷ The main problem to be addressed, in his eyes, was the gap between the interior-landscaping industry, with its horticultural knowledge, and the design profession – architects and interior designers, in particular – which, up until that point, had tried to develop plantscapes on its own. While interior landscapers understood plants' needs but had no real design skills, claimed Gaines, designers did not understand the principles of what he

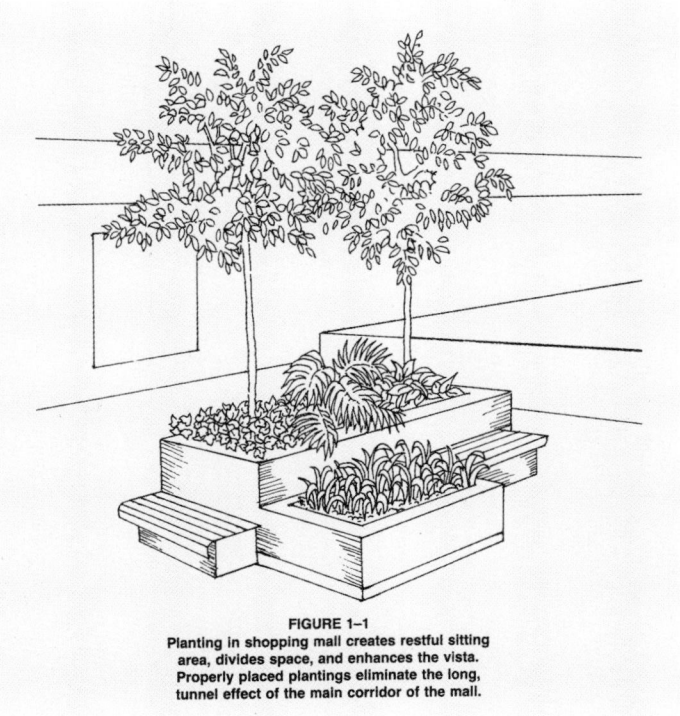

Fig. 100 Illustration advising on planting for a shopping mall, in Tok Furuta, *Interior Landscaping* (Reston, VA: Reston Publishing, 1983), fig. 1-1

called 'plant material'.[8] In order to meet the requirements of the rapidly expanding field, in which ever-larger financial investments were being made, he stressed the importance of collaboration. While books about domestic houseplants existed, Gaines explained, there was little that addressed the commercial sector.[9] He wanted his book to help the design profession understand how to work with foliage plants indoors, how to specify them, and how to ensure that adequate maintenance programmes were in place. He was adamant that the requirements for lighting, water supply, drainage, temperature and humidity needed to be included at the early design stage – a necessity that, in his view, had been overlooked in the past.

Gaines felt that as the pace of life increased, 'people need[ed] something to hold on to'. The use of indoor foliage plants was, therefore, not only fashionable but also necessary.[10] Beyond acting as visual screens, softening hard architectural surfaces, providing texture, articulating spatial volume, directing circulation and improving morale, Gaines firmly maintained that 'indoor foliage plants [were] symbolic of life itself and a denial of death'.[11] His words, which went to the core of the rationale for engaging with nature inside, touched on something that had been understood, albeit silently, for several centuries. While it was in plantscapers' interests to claim that plants were more than mere decoration in an interior

Fig. 101 Ruth Abbott, of the Denver supply firm Lehrer's Flowers, arranging azaleas planted around a restaurant in a shopping mall in Denver, 1986. Photograph by Lyn Alweis

setting, Gaines was one of a group who increasingly understood the importance of nature to human beings' very existence.

From 1976 onwards George H. Manaker taught a course in interior planting at Temple University in Philadelphia. In 1981 he published a textbook for his students in which he explained that much more training in the new field was needed. 'While advice was available for the "window sill" gardener', he explained, 'the student or practitioner of plantscaping ... requires a comprehensive knowledge of the interior environment and its manipulation.'[12] Manaker was fully aware of the important link between plantscaping and environmentalism. 'Interior planting is not a fad', he wrote. 'It is part of the back-to-earth, back-to-nature, back-to-the-senses movements evolving in culture today. Indoor plants may no longer be non-essential luxuries, but necessities just as our automotives and television sets are "necessary", playing a very important role in the American way of life.'[13] In 1983 Tok Furuta, another practitioner in the field, penned his own influential book directed at the plantscaping profession. It offered practical advice about how to position plants in offices and shopping malls to maximum effect (fig. 100).[14]

The plantscaping profession continued to expand in the United States through subsequent decades, much of it based near to the supply of plants

The Benefits of Nature Inside 167

in Florida and California. Planterra, established in 1973, and Plantscape House Inc., founded in 1980, were just two of the many commercial plantscaping operations formed at that time. By 1986 there were approximately 8,000 interior landscape companies in the United States, many of them one- or two-person operations, and about the same number again outside the country (fig. 101). Their work was made possible by the rapid development of new high-tech products, such as pressurised watering tanks and sub-irrigation systems.

In the first two decades of the twenty-first century, the number of plantscapers and designers who had taken on board the ideas articulated by Gaines, Manaker, Furuta and others increased dramatically worldwide. In Brighton, England, for example, the team working with the designer Oliver Heath have been employed across a range of interiors, from offices to hospitality spaces, education spaces, healthcare spaces, retail spaces and homes. Heath's projects included a striking installation at the Interface Showroom for Clerkenwell Design Week (2016), which featured a vertical planted wall. Working with nature inside forms part of Heath's wider approach towards designing for health and well-being, which, following the discoveries of the 1970s, also includes improving natural light, facilitating views out on to nature, incorporating natural materials, textures and patterns, and ventilating spaces.[15] In 2017 Heath was commissioned to create a meditation studio in London, named Re:Mind, which is linked to a retail space and a tea-tasting area. It contains a green wall designed by the British company Biotecture, and a rock-salt light installation believed to be able to remove toxins from the air (fig. 102). In order to persuade his clients of the benefits – human, environmental and economic – of engaging with nature inside, like the scientists working from the 1970s onwards, Heath backs up his work with rigorous research.

Heath fully embraces the concept of biophilia in the interior, a topic which has been discussed since the 1960s. The term was coined by the psychologist Erich Fromm in his 1964 book, *The Heart of Man*, in which he defined it as a 'psychological orientation of being attracted to all that is alive and vital'.[16] In the 1970s Conklin had been a strong advocate of humankind's inherent link to nature, publishing an influential article in the *American Nurseryman Magazine* entitled 'Man and Plants – A Primal Association' in 1972, in which he outlined the idea that people are genetically programmed to be near green, growing plants.[17] Two years later Conklin developed the same ideas in another article published in the same journal, and in 1978 he expressed his heartfelt belief that 'man is inherently unhappy in an environment in which there is an absence of plants and flowers'.[18] Conklin's intuitive approach was based on his childhood experiences of living on a farm. 'I knew the bay of the fox . . . I knew the inspiration of nature's foliage and flowers. It all seems lost now in our man-made cities', he explained regretfully.[19] On one level, Conklin was articulating a reason why people might employ him and his

Fig. 102 Oliver Heath Design, living wall in Re:Mind meditation centre, London. Photograph by Jonathan Bond, 2016

fellow plantscapers. On another, his words were part of a broader cultural movement that focused on human beings' need to remain close to nature.

In 1984 Edward O. Wilson published a book, *Biophilia*, in which, like Conklin before him, he articulated an evolutionary approach, defining biophilia as 'the innate tendency to focus on life and lifelike processes'.[20] Wilson was a biologist with a special interest in the behaviour of ants, and it was while he was undertaking fieldwork that he became aware of the calming and mysterious effect of the natural world, which, he claimed, 'is to be mastered but never (we hope) completely'.[21] Later in the book he admitted that 'we only think we have control', revealing himself to be one of the first in the field to recognise the agency of the natural world and the need for human beings to re-establish a balanced relationship with it.[22]

Wilson believed that human beings' affiliation with life included an innate aesthetic response to nature and an instinctive desire to recreate it in the form of landscapes and gardens. 'Some environments are indeed pleasant', he explained, 'for the same general reason that sugar is sweet, incest and cannibalism repulsive, and team sports exhilarating. Such response has its peculiar meaning rooted in the distant genetic past.'[23] Referencing the work of Leo Marx, he regretted the fact that humankind had followed the path of the machine, which, he felt, had driven a wedge between nature and culture

The Benefits of Nature Inside

and initiated man's destruction of the former. 'It is', he wrote evocatively, 'like burning a Renaissance painting to cook dinner'.[24]

For Wilson, the split between science and the humanities, and the dominance of the former, lay at the core of the problem. He was also dismissive about nature transformed into culture. 'Lawn grass, potted plants, caged parakeets, puppies and rubber snakes are not enough', he maintained.[25] Only nature in its wild state was acceptable to him. Along with Stephen Kellert, Wilson edited *The Biophilia Hypothesis*, a book of essays published in 1993. Covering a wide spectrum of subjects addressed from psychological, biological, cultural, symbolic and aesthetic perspectives, it set out to rationalise the innate and, in order to move away from what some considered romantic idealisation, to provide evidence for the existence of biophilia.[26] The hypothesis was now formulated as 'a human dependence on nature that extends far beyond the simple issues of material and physical sustenance to encompass as well the human craving for aesthetic, intellectual, cognitive, and even spiritual meaning and satisfaction'. Biophilia, it was suggested, was rooted in historic learning and persisted even in people who had lived in urban environments for several generations.[27] It was implied that the satisfaction of that craving led to psychological well-being, reduced stress levels, and promoted physical health.

From the mid-1970s onwards, bodies of urbanists, architects, landscape architects, interior designers and plantscapers began to ask serious questions about the benefits of plants in indoor spaces. Taking up the gauntlet that had been thrown down by Wilson and others – and in support of the work of the professionals involved, the industry that supplied the plants, and the institutions and commercial clients who invested in plants for their spaces – a number of environmental psychologists, clinical psychologists, sociologists, health specialists and others working in behavioural medicine began to develop theories about the effects of plants inside on human beings. Many of them expressed their ideas in evolutionary terms, claiming that 'the long evolutionary development of humankind in natural environments has left its mark on our species in the form of unlearned predispositions to pay attention and respond positively to the contents of those environments'.[28] They also believed that the restorative effects of plants were cross-cultural.

'Overload' and 'arousal' theories were two of the early justifications for bringing plants inside. Both are premised on the notion that 'environments with high levels of visual complexity, noise, intensity and movement' can be overwhelming but that restoration can be achieved and stress relieved by the calming effects of indoor plants.[29] The environmental psychologists Rachel and Stephen Kaplan, who were based at the University of Michigan, were awarded funding in the 1970s by the United States Forest Service to work on Michigan's wilderness. Interested in the effect of nature on people's relationships and health they developed what they dubbed 'Attention Restoration

Fig. 103 Potted philodendrons decorate an open-plan 'landscaped' office in Hamburg, 1974. Photographer unknown

Therapy'. Rooted in the belief that nature had 'high fascination value', it provided, they maintained, a remedy for mental fatigue.

Roger S. Ulrich's main interest also lay in the links between interior environments and health. Although his background was in the social sciences, he moved into the area of evidence-based design. After a period spent in Sweden, in 1988 he became a professor of architecture at Texas A&M University, where he helped to establish the Center for Health Systems and Design. Although he also adopted an evolutionary approach, unlike the Kaplans, Ulrich maintained that human beings' responses to natural environments were not registered consciously but were, rather, 'quick on-set affective or emotional reactions'.[30] His work was linked to studies of the emotions being undertaken at the time, as well as to contemporary developments in neurobiology.

Through the final two decades of the twentieth century, and the first decade of the twenty-first, many other researchers measured pulses, heartbeats, blood pressure, muscle contractions and immunoglobulin levels, and undertook brain scans, in their attempts to provide scientific evidence of human beings' responses to the natural world, both outside and inside buildings. The work was conducted in the United States, the

Fig. 104 A variegated dracaena *(front)*, philodendron *(rear)* and other indoor plants in a container at the landscaped office of the Swiss Banking Corporation in Basel, 1970s. Photographer unknown

Fig. 105 Potted plants and cut flowers help enliven an open-plan office, 1970s. Photographer unknown

United Kingdom, Norway, Sweden, Japan, the Netherlands and China, among other countries, and the results were published in global academic journals.[31] Hard evidence was needed not only to justify the funding that went into the research, but also to reassure all those who wanted to invest in plantscaping that it paid for itself, either indirectly by creating efficiencies, or directly by resulting in financial profit.

The experiments focused on a range of specific interiors – including offices, hospitals and prisons, and, a little later, hospitality and retail spaces – in which well-being was an important issue. Since the 1960s plants had played an important role in what was described at that time as the landscaped office, or *Bürolandschaft*, a concept developed by a German space-planning firm called Quickborner (figs 103–5). The firm was owned by the brothers Wolfgang and Eberhard Schnelle, who had set out to create an office system that functioned on a human scale and was more flexible than the hierarchically arranged rows of desks that had previously been the norm.[32] A range of plants, including dracaenas, philodendrons, yuccas, weeping figs, Kentia palms, umbrella trees, fatsias, rubber plants, Swiss cheese plants, ivies and bamboos were introduced into landscaped offices. They served multiple purposes, including absorbing noise, replacing partitions, providing a level of privacy and acting as directional signs, and played several aesthetic roles, among them creating focal points, filling voids, and adding colour, texture and warmth.

In the early 1960s the German project came to the notice of an American researcher named Robert Propst, who was working along similar lines for the Herman Miller company in the United States. In 1964 Herman Miller launched a new concept called 'Action Office', on which Propst collaborated with the designer George Nelson. The first Quickborner office landscape in America was created for Dupont in Wilmington, Delaware, in 1967.[33] Subsequently, plants played an important role in numerous office projects across the globe that emulated the work undertaken in Germany and the United States (fig. 106).

Serious research on the effects of plants in offices began in the early 1970s. By 1978 Conklin was able to claim that 'attitudes in the planted office seem to indicate individual and collective morale very much improved and absenteeism considerably down'.[34] At that time the energy crisis was forcing builders to turn to more efficient products for construction, and office buildings were being hermetically sealed. The new materials that they used contained toxins that led to a multiple-symptom disorder in office workers known as 'sick building syndrome'. As a result, in addition to the work being done on plants as stress relievers, research was also undertaken to address their capacity to combat air pollution.

Research on the effects of plants on office workers continued into the 1990s. In 1993 Rachel Kaplan reworked her ideas about 'nearby nature' and 'mental fatigue'.[35] Her rather vague conclusion was that: 'To be able to glance up from one's work and experience bits of nature is likely to be

Fig. 106 Plants in an open-plan office in Helsinki, designed by Martela Oyj, 1980s. Photograph by Studio 13

helpful.'[36] Five years later a group of Norwegian researchers described the result of a study that had asked workers twelve questions relating to their health. From their responses the researchers reached the conclusion that 'complaints regarding coughs and fatigue were reduced by thirty-seven and thirty per cent respectively if the office contained plants'.[37] By the early twenty-first century the computer had become a key feature of the office and, in 2015, using stress-recovery and attention-restoration theories to analyse their findings, another group of Norwegian researchers reported that 'participants who had plants present at the workstation showed an improvement in their performance of a directed attention task over time, while participants who sat in a lean control environment did not'.[38]

A 1976 paper addressed the role of plants inside a psychiatric hospital.[39] Two years later, Conklin described an experiment that a group of researchers had undertaken at the Dunlap Psychiatric Hospital in New York City, in which a number of patients had been exposed to flowering chrysanthemum plants that had been casually placed on several dining tables, while the other tables in the room had nothing on them. The research concluded that eleven per cent of the people sitting at a table with a plant on it had eaten more than the others; twenty-one per cent had spent more time at the table; and thirty-three per cent had conversed more with their fellow patients. That was taken, rather simplistically perhaps, as proof of the benefits of indoor planting. The sophistication of experiments of that kind was to increase over the years, however, as more probing questions were asked, controls were added, and the test results were analysed more rigorously.

In 1984 Ulrich had suggested that the sight of nature outside hospital wards could help alleviate anxiety or stress.[40] Over two decades later the researchers Seong-Hyun Park and Richard H. Mattson took the work one step further by testing the medical and psychological effects of plants and flowers that had been introduced into the hospital rooms of post-operative female patients. Patients with plants in their rooms were found to have had 'shorter hospitalisations, less need for analgesics, lower ratings of pain, anxiety and fatigue, and more positive feelings', compared to those without plants.[41] The work claimed to prove what the Victorians, who had filled their hospital wards with flowers, had undoubtedly known intuitively (fig. 107). The Khoo Teck Puat hospital, which opened in Singapore in 2010, is surrounded by plants inside and out, while the Crown Sky Garden in the Ann & Robert H. Lurie Children's Hospital in Chicago, opened in 2012, also aims to have a relaxing effect. Within a glass greenhouse perched high up on the eleventh floor of the building, the designer, Mikyoung Kim, combined light, water and colour with bamboo groves, marble fountains, natural stone and reclaimed wood.[42] The same approach was taken to childcare facilities, in which plants were seen to enhance the experiences of the children who inhabited them. The privately run John Hancock Child Care Center in Boston in the United States, which opened in 1990, contained a densely planted interior garden.

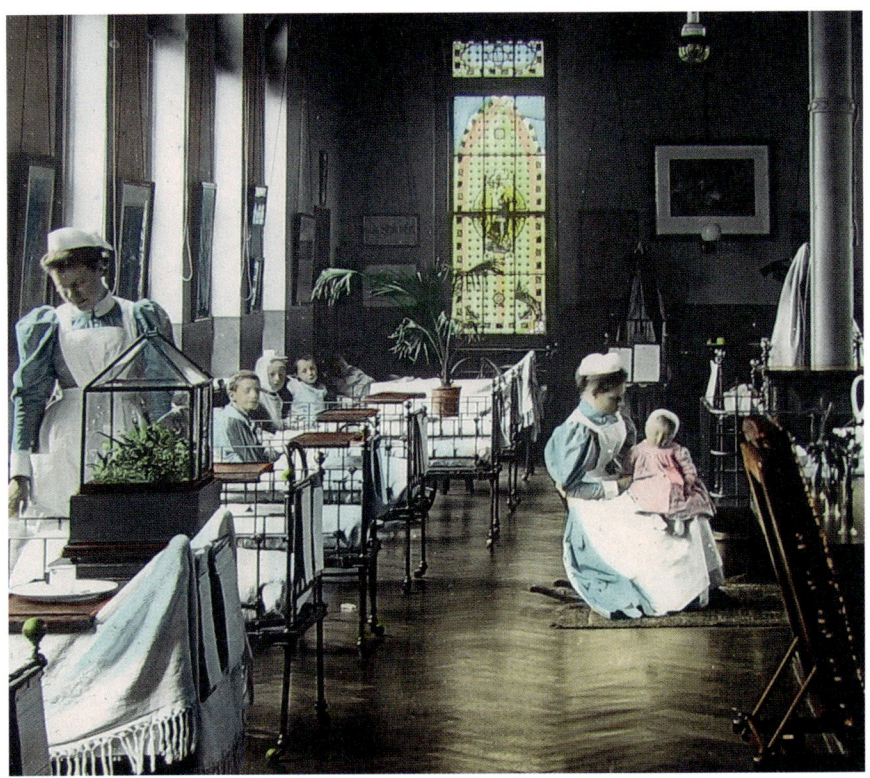

Fig. 107 Hand-tinted photograph showing a children's hospital ward with plants in pots and cases, early 1900s

The 1980s saw similar work being undertaken in a range of other institutions. However, although many plant-filled hotels emerged in that period, little was published about the role and impact of plantscapes in buildings serving the hospitality business. One exception was a short piece penned in 1992 by Michael R. Evans, the former Associate Professor of Hotel, Restaurant and Institutional Management at the Virginia Polytechnic Institute and State University. Evans provided a case study of the Opryland Hotel in Nashville, Tennessee, suggesting that the investment in indoor planting made over several years had proven financially successful.[43] Following the precedent set by John Portman's Hyatt hotel chain, the Opryland Hotel opened in 1977 as the twelfth-biggest hotel in the United States. The research maintained that the hotel's plantscaping encouraged guests to return and resulted in an occupancy rate well over the national average. By 1983 the hotel's earliest plantscaping scheme, the first of three indoor gardens, was in place. It contains a two-acre conservatory resembling iron-and-glass examples from the Victorian era, which features a decorative, 72-foot-high 'Crystal Gazebo' fountain at its centre, as well as walkways on which guests can immerse themselves in lush foliage and a semi-tropical

Fig. 108 Trees and plants in the domed glass Delta Atrium at the Opryland Hotel in Nashville, designed by Earl Swensson Associates, part of the phase IV expansion completed 1996. Photograph by Greg Balfour Evans

ambiance. Most of the plants were acquired from a Florida nursery in 1982, many of them grown especially for the hotel. Five years later, the one-and-a-half-acre Cascades Atrium was constructed. As well as including large numbers of plants and flowers, it also contains a waterfall, a koi pond and a fountain. A little later, the area known as the Delta Atrium was added (fig. 108), featuring an indoor river along which guests can take a boat ride, and a wide range of subtropical trees and plants, including magnolias, gardenias and camellias. By 1992 the hotel's twelve acres of indoor space contained about 18,000 plants representing around 600 species, including fig trees, banana trees and a small number of miniature orange trees. By 2011 there were sixty-three varieties of palm, among them a 70-foot-tall sugar palm that the horticulture manager of the hotel, Hollis Malone, had acquired from a collector in Miami in 1986.[44] The annual horticulture maintenance budget was about 1.2 million dollars, and a staff of fifty-two was needed to tend the plants.[45] A drip irrigation system was installed, and plantscaping was used to justify higher prices for the rooms that overlooked the atrium.

Research relating to plants in indoor retail environments came a little later. In 2016 the researchers Sigal Tifferet and Iris Vilnai-Yavetz used

The Benefits of Nature Inside

evolutionary theory, once again, to assess the effects on shoppers of placing a plant in a retail environment, as opposed to having a vase or nothing at all in the same space.[46] The authors concluded that 'the plant was superior to both no stimulus and the vase in increasing perceived service quality and customer satisfaction'.[47]

Complementing work on the effects of indoor plants on human beings' health and well-being, the role of indoor plants in tackling air pollution was also widely addressed from the 1970s onwards. The subject came to a head in the wake of a 1989 report by Bill Wolverton, who, by that time, had been a senior research scientist at NASA's Stennis Space Center in Mississippi for eighteen years. Back in 1973, NASA had discovered that Sky Lab 3 had been contaminated by more than 300 volatile organic chemicals (VOCs), and Wolverton had been called in to investigate the problem.

By the early 1980s Wolverton had become renowned for describing plants as 'the lungs of the earth'. He had discovered that certain species – peace lilies, areca palms, lady palms, fig trees and the golden pothos among them – were more effective air purifiers than others. In 1984 he published some of his early research and, three years later, he and his team set out on a two-year project, jointly funded by NASA and the Associated Landscape Contractors of America, to evaluate the ability of twelve common houseplants to remove chemicals from sealed units. Wolverton's 1989 report on the use of interior landscape plants for abating indoor air pollution, based on the above project, was published by NASA. It began by outlining sick building syndrome in general and went on to focus on the air-pollution problems that NASA had identified in sealed space habitats. The main trio of toxins that needed eradication were understood to be benzene, trichloroethylene and formaldehyde, and the decision to use plants and soil to remove them from the environment seemed obvious to Wolverton. 'Since man's existence on Earth depends upon a life support system involving an intricate relationship with plants and their associated microorganisms', he wrote, 'it should be obvious that when he attempts to isolate himself in tightly sealed buildings away from this ecological system, problems will arise.'[48] The bulk of the report describes the tests that were performed and the main result, which was that 'low-light-requiring houseplants with activated carbon plant filters, have demonstrated the potential for improving indoor air quality by removing trace organic pollutants from the air in energy-efficient buildings'.[49]

That simple statement had far-reaching effects. Wolverton retired from NASA in 1990 but was retained as a consultant. His clean-air studies provided the interior landscape industry with a persuasive marketing strategy: they could now sell plants not just as decorative items, but also as a solution to sick building syndrome. In 1996 Wolverton published a popular book that set out to educate the public about the '50 indoor plants that purify the air in homes and offices'.[50] Other researchers in academic institutions as far apart as Sydney, Australia, and Lancashire, England, have carried on the scientific work that Wolverton initiated with NASA.

After a slight decline in the 1990s, the plantscaping industry bounced back in the 2000s as the commercial sector came to understand the proven benefits of indoor plants, and as engaging with them became increasingly fashionable (just as it had in the nineteenth century). The countless studies indicating the positive effects of nature inside on people inhabiting a wide range of spaces impacted on both private individuals and the commercial sector alike. In retail environments there was an obvious effect on profits, and if efficiencies could be achieved in hospitals and prisons, it was reasoned, funding could be more effectively deployed.

The experiments that were undertaken from the 1970s onwards demonstrated that human beings had positive psychological and physical responses to indoor plants and that nature was important in maintaining a healthy indoor environment. Victorian advisors on window gardening had known full well that plants could bring happiness to a home, cheer up the sick and lonely, and have a therapeutic effect on invalids. In addition, they had been aware that plants had a physical effect on the indoor environment and its inhabitants, providing air purification, humidification and temperature control. By the twenty-first century the benefits of plants inside were believed to be even more extensive. By absorbing and dulling noise, for example, indoor plants were understood to improve the acoustics in noisy offices or on busy streets.

Whatever the validity of all these experiments, it is clear that, in the first decades of the twenty-first century, large numbers of people across the world were convinced by them. Nature inside was more popular than ever before, both in people's homes and in public spaces. As a consequence, Conklin and his colleagues were in unprecedented demand. The work undertaken by Wolverton and others in the area of air pollution was just one part of a large body of research that influenced the activities of environmentalists, for whom nature inside held the promise of a new relationship between human beings and the natural world.

CHAPTER 10

Greening the Interior

The 1990s promise to be the decade of the environment.[1]
Charles A. Lewis

Alongside the research that was being undertaken from the 1970s onwards into the psychological, physical and environmental benefits of indoor plants and flowers, popular awareness was growing about the worrying effects of the long-term imbalance in the relationship between human beings and nature. One of the first calls for a global environmental movement that recognised the damage human beings had caused came in 1962 with the publication, in the United States, of Rachel Carson's environmental science book, *Silent Spring*. In its introduction, Lord Shackleton described the book's subject as 'ecology, or the relation of plants and animals to their environment and to one another'.[2] It had come to Carson's notice that many species were being destroyed by the presence of toxic chemicals in the environment, and she strongly encouraged her readers to become aware of 'The science of ecology [which] teaches us that we have to understand the interaction of all living things in the environment in which we live.'[3] The book provided an early warning of what happens when human beings start tampering with forests and removing the 'weeds' upon which insects and animals depend.[4]

Carson's book stimulated the emergence of a popular environmentalist movement, which, by the early 2000s, had come to embrace a broad range of related issues and causes. In 2007 Timothy Morton defined environmentalism as 'a set of cultural and political responses to a crisis of human relationship with their surroundings . . . [that] could be scientific, activist, or artistic, or a mixture of all three'.[5] Arguably, by bringing nature inside both private and public spaces, the interventions that architects, interior designers, plantscapers and home-makers were making into interiors across the globe constituted an aspect of the artistic face of environmentalism. Like all causes, however,

Fig. 109 Desert and semi-desert plants inside one of the biodomes at the Eden Project in Bodelva, Cornwall, designed by Nicholas Grimshaw, opened 2001; photographed 2016

environmentalism was open to abuse. The term 'greenwashing' was coined by sceptics to describe the exploitative nature of much of the marketing that has been developed around certain goods and services to give commercial companies green credentials. The term greenwashing implies a level of duplicity and, viewed from that perspective, public-sphere green interiors could be seen in the same light. While, on the one hand, they are directly beneficial to human beings on several levels, on the other, they encourage people to work harder and spend more money in order to keep the capitalist economy moving.

By the early twenty-first century nature inside had acquired a set of new symbolic meanings. Over a period of about a century and a half, a shift had taken place from nature being brought inside as a form of compensation for its separation from human beings as a result of urbanisation, to an almost obsessive inclusion of greenery indoors linked to the environmental crisis of the early 2000s. A tension developed, however, between nature inside signifying a genuine commitment to environmentalism, and its providing a way in which human beings could maintain control over the natural world. While projects such as the Biodome in Montreal, the Eden Project

Fig. 110 Palms, ferns, and other plants inside the Temperate House at the Royal Botanic Gardens, Kew, opened 1852; restored 2018. Photograph by Marshall Black, 2018

in Cornwall (fig. 109) and, most recently, the restored Temperate House in Kew Gardens (fig. 110) aimed to educate people about plants from around the world and the importance of biodiversity, they also revived the idea of imprisoning plants and flowers under glass in the manner of the colonial glasshouses of the past.

As explained in Chapter 9, in order to persuade the commercial sector to use plants inside, and to support the industries that facilitated the supply and installation of indoor nature, the beneficial effects of plants inside a range of public and semi-public spaces needed to be evidenced. In the popular arena of the home, however, the appeal of nature inside continued to be more emotional than rational. Over a century earlier the authors of advice books had widely recognised that the attraction of indoor nature was almost exclusively to the heart rather than the head.[6] John R. Mollinson had written in 1877 that: 'The love of flowers arouses within us all the kindliest feelings of our nature, humanises the heart, and fills the mind with pleasant thoughts and associations.'[7] In the commercial context, too, indoor nature could appeal to humans' affective side by helping to create a multisensory experience, as John Portman had recognised by the late 1960s.[8] Alongside

Greening the Interior 183

the effects of lighting and colour in nineteenth-century interior décor, plants had added a softening element, texture, scents and a sense of exoticism and luxury, and, as Shirley Hibberd had recognised, offered 'rest, solace and refreshment'.[9] Indoor greenery also evoked the hot and humid atmosphere of the jungle. Over a century later, Morton suggested that a sensorial relationship with the natural world was still possible, claiming that 'Nature compels feelings and beliefs'.[10]

That emotional relationship with plants and flowers came to the fore in the 1970s British domestic interior. Like the Victorians before them, many urban and suburban residents brought plants and flowers inside as part of a general trend in interior decoration that looked back to a set of idealised and romanticised values linked to life in the pre-industrial era. The actions of professional and amateur home-makers were driven both by a renewed fear of losing humankind's connection with the natural world and with its pre-industrial, pre-urban roots, and by a new realisation that human beings' long-term exploitation of nature was resulting in serious environmental problems. One response to those circumstances was widespread nostalgia. While some of the same impulses that had been expressed in the Victorian era resurfaced, from the 1970s onwards they existed within a dramatically transformed context that had seen the rise of – and growing disillusionment with – architectural modernism, unprecedented advances in technology, the emergence of a fully formed global economy and system of communications, and an acute awareness of environmental issues.[11]

In 1970s Britain that sense of nostalgia was partly manifested in a revival of interest in the historical English country house and its interior décor. Efforts were made to replicate both its spirit and its aesthetic in contemporary urban and suburban middle-class homes. The exhibition *Destruction of the Country House, 1875–1975*, held in London's Victoria and Albert Museum in 1974, stimulated that trend. In addition to linking people to an eighteenth- and nineteenth-century past in which ideals of Britishness and the empire were still intact, the return to this domestic style also offered a vision of a simple life free from the complications of advanced technology, and provided domestic comfort and a sense of belonging.

Through the 1970s, 1980s and 1990s this new nostalgic language of interior decoration took hold in popular middle-class urban and suburban settings. Emulating the interiors of the eighteenth-century nobility, it was characterised materially by festooned curtains, chandeliers, comfy sofas, bookshelves, muted colours, antique furniture, chintz, and decorative use of plants and flowers. In 1977 the advice-book writer Ann Bonar explained that:

> Hanging and climbing plants, such as ivies, rhoicissus and tradescantia can be used as room dividers or put on bookshelves, or placed in troughs on the floor and trained to climb up canes to the ceiling. You can train cissus and philodendron as living decorations round mirrors or up and along window frames.[12]

Fig. 111 The front cover of Yvonne Rees, *Essential English Country Style* (London: Ward Lock, 1995)

The taste for neo-Victorianism of those years is perhaps epitomised in her statement: 'Aspidistras have come into their own again.'[13]

Domestic bathrooms were filled with plants in hanging macramé baskets, resembling those that had held orchids in the nineteenth century. The Swiss cheese plant also featured strongly once again. Bonar described its aesthetic contribution:

> Originally from tropical monsoon forests, it has really large leaves, deeply slashed at the edges. It also has holes in its leaves. When the lights shine through these and cast shadows on to a white wall, the combination of silhouette and the plant itself is extremely effective.[14]

Ferns and palms also regained popularity, as did rubber plants and weeping figs.[15] That nostalgic trend remained visible up to the turn of the century, and a 1995 publication entitled *Essential English Country Style*, written by Yvonne Rees (fig. 111), depicted a simple lifestyle expressed through pine dressers, rocking chairs, blue-and-white china, pewter plates, and traditional kitchen items, such as brown ceramic jelly moulds. The dominant characteristic of that simple life was its appeal to the senses, evoked, for example, by the smells of 'sweet woodsmoke, a rose and lavender-based pot pourri, heady beeswax polish, newly-mown grass and a damp spaniel'.[16] Bouquets of country flowers, combined with greenery and roughly displayed in rustic pots, were part of that sensorial mix.

Greening the Interior

Victorian-style domestic conservatories also made a reappearance. Descriptions of them focused on their atmospheric qualities and their appeal to the senses. Mary Gilliatt wrote evocatively that, for her, 'The warm, damp smell of these graceful glass constructions has always been a heady one.'[17] Because the advice books of the era were intended to appeal to home-makers, rather than corporate-finance decision-makers, the language used by Gilliatt and her advice book peers differed dramatically from that adopted by the environmental psychologists who were trying to prove the effects of plants and flowers on human beings by scientific means. The 1980s saw similar advice books emerge in France, Sweden, Ireland and the United States, among other countries, suggesting that Britain was not alone in its promotion of an interior decorating style that reflected a nostalgic longing for a past era. Like their nineteenth-century antecedents, the books constituted an industry in themselves, one that built on growing popular disillusionment with the idea of a future that was rooted solely in the promise of new technology.

By the dawn of the twenty-first century, the nostalgia for nineteenth-century domestic styles had largely been replaced by an infatuation with what has come to be called 'mid-century modernism', the popular domestic interior decorating idiom of the 1950s, which, as was shown in Chapter 6, had also warmly embraced nature inside. This was a different kind of nostalgia, a retro-futurism that embraced a past vision of the future rather than a past vision of the past.

By the early twenty-first century, plants (rather more than flowers) had also become part of a global late-modern movement in residential architecture that revived several strategies tried and tested in the first phase of inter-war, and immediate post-war, international modernism. In crowded cities around the world, a lack of outside spaces around houses and apartments drove people to bring nature inside once again. The populations of several Asian countries maintain long-standing relationships with nature through their allegiances to a variety of religious philosophies: Shintoism, Buddhism, Hinduism, Taoism, Confucianism and others. Those relationships are complex and variable, however, and are often linked to specific locations and cultures. According to Ole Bruun and Arne Kalland, for example, 'Whereas sandal trees have been regarded as sacred and inhabited by spirits on Java and to a lesser extent on Sumba until recently, there exists no indication that people on Timor ever held such beliefs.'[18] The same authors also made the more general point that: 'The nature–culture distinction may be less categorical in the East than in the West and [that] the Japanese do not distinguish sharply between nature and culture.'[19]

Added to the fact that, by the twenty-first century, Japan was a country in which urban space was at a premium, and external gardens and direct sunlight were considered luxuries, this historical closeness between nature and culture may have prompted the numerous progressive Japanese architects who adopted a globalised late-modern idiom that embraced

Fig. 112 Inside a full-scale reconstruction of the Moriyama House, Tokyo, designed by Ryue Nishizawa, 2005, recreated for the exhibition *The Japanese House: Architecture and Life after 1945* at the Barbican, London, 2017. Photograph by Veronika Lukasova

plants inside. As well as reworking indigenous traditions, Japanese architects and designers integrated Western ideas into their buildings, especially those linked to modernism, and introduced plants inside as a means of softening and humanising their indoor spaces. In 2005 Ryue Nishizawa of Studio SANAA designed a house for Yasuo Moriyama in Tokyo. It comprised a number of variously sized cubes, the spaces both within and between the cubes containing numerous plants (fig. 112). In 2009 the architectural group Suppose Design, which is based in Tokyo and Hiroshima, designed a house in Nagoya that featured an indoor garden.[20] An architectural critic described Suppose Design's spaces as 'otherworldly . . . almost how one might imagine a greenhouse within a space ship, designed to keep the passengers healthy, happy and sane'.[21] UID Architects also placed a tree inside a house, called Shrimp, which was built in Fukuyama in 2014. As it had decades earlier in Ludwig Mies van der Rohe's Villa Tugendhat (1929–30), indoor vegetation in these Japanese houses served to provide a counterpoint to their otherwise hard, geometric architectural structures and industrial materials. They also brought nature into the city.

Architects and designers working in a range of other Asian countries also filled their indoor spaces with plants. In a 2018 house designed by Aaksen Responsible Architecture in Bandung, Indonesia, a Brazilian fire-tree pierced

Greening the Interior

Fig. 113 The indoor garden at G-Tower, Incheon, designed by HAEAHN Architecture, completed 2013. Photograph by Inigo Bujedo Aguirre, 2014

the white zinc-coated roof of the building from within. Branches protruded through circular openings cut into the decks of a house designed by Budo Pradono in Jakarta (2015), while trees emerged through openings in the grass-covered roof of another of that studio's projects in the city of Depok (2011). Chang Architects' 2005 Elok House in Singapore also contained trees, which wound their way through its internal spaces and created a cool microclimate.[22] Although not residential, the G-Tower in Incheon, South Korea, designed in 2013 by HAEAHN and containing restaurants, banks and a post office, features an indoor garden (fig. 113).

Demonstrating the global reach of indoor nature in domestic settings in the early twenty-first century, similar strategies were also implemented on the American continent. In Mexico, for example, a forest of sweet gum trees emerged through openings in the concrete slabs of the Julio Hernández House in Morelia, designed in 2016 by Roof Arquitectos, while in 2017 the architect David Guerra created a house in south-eastern Brazil that he filled with tropical palm trees and colourful plants.[23] The desire to include nature inside in those locations could be explained by the fact that Asia and Central/South America were late adopters of the Western model of architectural modernism. However, seen from a less Western-centric perspective, they could also be seen to be reasserting an aspect of their own local cultures, that is, a closer alignment between nature and culture than existed in the West. Also, as many of the exotic plants used in the West from the eighteenth century onwards had originated in Asia and the Americas, local architects and designers had easy access to them. Most of the plants used by American plantscapers in the late twentieth and early twenty-first centuries were imported from Mexico, Costa Rica and Colombia, and the expansion of their work encouraged the growth of large global businesses in those countries.

A number of residential apartment blocks in several urban centres across the globe were also given the green treatment. In Europe, Milan's Bosco Verticale ('Vertical Forest'; fig. 114), a pair of towers designed by Boeri Studio and constructed over the period 2009–14, features large balconies that host full-size trees, shrubs and ground cover, giving the exterior of the building a wooded appearance. While there are practical benefits, such as noise abatement and a means of dealing with smog, the main effect of the towers on Milan's skyline is aesthetic. Boeri applied the same principle in his proposal for the Trudo Vertical Forest social-housing project in Eindhoven in the Netherlands. Across the Atlantic, in Toronto, a similar tower has been proposed by the architectural firm Penda.[24]

While, in the years after 1990, plants and flowers featured widely in the interior décor of many private homes across the world – both nostalgic and contemporary – the same period saw plants and flowers continuing to penetrate global public and semi-public buildings. Countless immersive, jungle-like public atmospheres were created, which facilitated the possibility of escape from the stresses of city life, encouraged an engagement with

Fig. 114 The Bosco Verticale, Milan, in winter, designed by Boeri Studio, completed 2014. Photograph by Mairo Cinquetti, 2019

commerce, provided added value for the owners of the buildings in question, and, depending on local legislation, could allow developers to construct larger buildings than would otherwise have been possible. In the form of enclosed trees, indoor gardens, carefully positioned individual plants and groups of plants, living walls, or a combination of several of those plantscaping strategies, nature infiltrated scores of offices, large hotels, corporate buildings, railway stations and airports, shopping malls, retail outlets and other commercial spaces, well-being and ecology centres, and cultural institutions. Although tactics used by the inter-war and post-war twentieth-century modernists were often revived, the scale was augmented significantly and nature inside took on new meanings.

The agendas underpinning the decisions to introduce nature into those large public interior spaces ranged from a genuine attempt to improve people's lives to unapologetic commercial exploitation. At both ends of the spectrum, plants' functional qualities – their capacity for air cleansing and oxygenation, and their links to well-being – were used to justify their presence. However, for the majority of the population who experienced indoor plants, it was undoubtedly their emotional, sensorial, aesthetic and atmospheric effects that were most important.

Fig. 115 Palm trees growing in the Winter Garden Atrium in the World Financial Center (Brookfield Place), New York, designed by César Pelli with landscape designer Diana Balmori, opened 1988; reconstructed 2002. Photograph by the author, 2016

As they had in the late nineteenth century, plants and flowers transferred domestic values into the public arena such that the boundaries between the private and public spheres became increasingly porous. As an adjunct to new architectural typologies and interior design strategies that fostered ambiguous spaces, the inclusion of nature in urban indoor settings provided an important way of encouraging seamless living. While new personal communications and information technologies permitted previously limited activities to be undertaken in a wide variety of environments – laptops and mobile phones, for example, are extensively used in coffee shops, book shops and airport lounges, among many other places – indoor planting schemes reinforced that new-found flexibility and freedom.

The familiar strategy of either leaving trees on site, or bringing them inside, which had characterised buildings by Frank Lloyd Wright and Le Corbusier and had featured in New York's Four Seasons Restaurant in earlier decades, re-emerged after 1970 on a significant scale. Palm trees were, as ever, particularly popular. César Pelli's atrium at New York's Battery Park in the World Financial Center (1988) was among the first large urban interior public spaces, and one of the most spectacular, to feature palm trees (fig. 115). It contained a grove of sixteen thread palms, each 45 feet tall, which had been transported to New York from California. As Martin Gottlieb explained in the *New York Times*, that particular species was 'chosen for its monumental quality, lushness and elegant trunk'.[25] Ten storeys high, the atrium was designed by Pelli in collaboration with his wife, the landscape designer and environmentalist Diana Balmori, who created the winter garden. The terrorist attacks of 11 September 2001 seriously damaged the site, but the original 1988 space was reconstructed in the following year. Although linked to the Financial Center, the space is dedicated to culture, hosting concerts, ballets and art exhibitions. The drama of the towering palm trees reinforces the immersive and emotive capacity of nature inside, as well as animating the otherwise static, late-modern steel-and-glass architecture that encases them. Some years after the atrium opened, the original palms grew too tall for the space and had to be replaced with smaller ones.

In Europe, the vast glass dome of Basel-Mulhouse Airport's Skyview Lounge (2005) also covers a number of exotic palms and a calming pool of water, introduced to alleviate the stress of the anxious travellers who pass time there.[26] The Knockbreda health centre in Belfast (2008) also features a palm tree in its central, naturally lit atrium. The aim is to provide a life-affirming and soothing waiting area for patients. Other trees were also brought into buildings to great effect in the first decades of the twenty-first century. As mentioned in the Introduction, in 2011 the wedding of Prince William to Catherine Middleton in Westminster Abbey, London, was enhanced by the addition of four tons of foliage. The aim was to recreate a 'lavish English country garden'.[27] It was an example of pure theatre, choreographed by the event's 'floral artistic director', Shane Connolly.[28]

Norman Foster + Partners have introduced trees into several Apple stores across the globe (fig. 116). The Singapore store, opened in 2017, was built on Apple Orchard Road, a former site of fruit and nutmeg orchards, which inspired the store's green theme. Singapore's status as the greenest place in Asia also influenced the design. Eight mature trees create a shaded 'green orchard' outside, while, inside, twelve weeping fig trees form what Apple calls its 'Genius Grove', a replacement for the 'Genius Bars' of earlier stores. It was included, the Foster office claimed, not only for the strong visual aesthetic but also to 'stimulate all human senses – enriching the temperature, the smell, and the aural qualities of the space. It creates an oasis of tranquillity in the busiest part of the store.'[29] The doughnut-shaped planters double as seats for customers.

In addition to the dramatic inclusion of full-size trees in indoor settings, many of the most powerful and memorable plantscapes that appeared in the decades around the turn of the century were large indoor gardens. Madrid's Atocha railway station, which was reconstructed in 1992, contains a full-scale botanical garden and turtle sanctuary. Over 7,000 plants and trees, representing over 260 species, cover an area of 13,000 feet. A swamp, filled with Malabar chestnut trees, occupies one end. Foster + Partners'

Fig. 116 Weeping fig trees are used to make indoor groves at the Apple store on Regent Street, London, designed by Foster + Partners, completed 2016. Photograph by Alena Kravchenko, 2017

Fig. 117 The indoor roof garden at Crossrail Place, Canary Wharf, London, designed by Foster + Partners, completed and photographed 2015. Photograph by Jim Dyson

Fig. 118 Looking out from the Sky Garden at the top of 20 Fenchurch Street, London, designed by Gillespies, completed 2011. Photograph by Steve Tulley, 2015

Crossrail terminal in London (2015; fig. 117) and Oslo Central railway station also use large installations of indoor greenery to calm those who pass through them.

The planting that the landscape architectural practice Gillespies developed for London's Sky Garden, a commercial space that opened in 2015, was dominated by drought-resistant Mediterranean and African species (fig. 118).[30] Aimed at a general audience, Sky Garden is a leisure space at the top of a commercial skyscraper located at 20 Fenchurch Street, to which the public is invited (albeit with a number of restrictions) to eat, drink, wander around, take photographs and generally enjoy themselves. Visitors are wooed there by the spectacle of the planting, lulled into relaxation by the soothing green surroundings, and then invited to spend money at the venue's many eateries. The Sky Garden recalls the leisure spaces of the nineteenth century – the Crystal Palace, the Royal Aquarium and Summer and Winter Gardens, and the Alexandra Palace among them – which used indoor plants to encourage leisured consumption. The Fenchurch Street building has the extra advantage of a captive audience in the people who work in it. As the architectural critic Rowan Moore pointed out, however, the Sky Garden is not a park in which children can run around and play. Rather, the reason for its existence, Moore claimed, was that it enabled its developers to gain planning permission.[31]

Fig. 119 Green wall created by Biotecture for the Anthropologie store on Regent Street, London, completed and photographed 2009. Photograph by Jeff Gilbert

Fig. 120 Forest Valley and Rain Vortex waterfall at Jewel Changi Airport, Singapore, designed by Peter Walker and Partners, completed and photographed 2019. Photograph by Suhaimi Abdullah

In the early years of the twenty-first century plants and flowers were also used on a smaller scale in many other public buildings, airports and malls among them. As well as encouraging shoppers to consume, they acted as directional signs, indicated changes of level, softened hard architecture, and added colour and texture. The designers of the Swedish fashion store & Other Stories, for instance, placed small plants – cacti and succulents – at various points in the store to give its customers a sense of being at home, to provide warmth and colour, and to appeal to its young audience, for whom owning such greenery constituted a grounding lifestyle choice in an age of uncertainty.

In its European flagship outlet in London's Regent Street, the American fashion store Anthropologie invested heavily in 2009 in what are interchangeably called green, living or vertical walls, one of the latest forms of indoor planting (fig. 119). The brainchild of a French botanist named Patrick Blanc, who created his first living wall in 1986, they were soon visible on both the exteriors and interiors of buildings. A dramatic exterior example was added to Jean Nouvel's Musée du quai Branly in Paris in 2004.

Living walls consist of plants rooted in fibrous material, with the water that hydrates them being reused and/or recirculated. Some are pre-grown, while others are planted on site. They can be found in a variety of usually commercial indoor spaces, particularly corporate offices, such as the Tryg Insurance headquarters in Copenhagen (constructed over the period 2007–11), an office in Wilson Street, London (2013), and the David Rubenstein Atrium in New York's Lincoln Center (2009). The wall in the Wilson Street office comprises ferns, ivy, peace lilies and begonias, and was designed by Biotecture, the supplier of the wall in the Anthropologie store, and in Oliver Heath's design for the Re:Mind project.[32] Again with the aim of calming stressed passengers, Gate 25 at Heathrow's Terminal 3 features what it calls a 'Garden Gate', another Biotecture living wall, composed of 1,680 plants including ivy and lilies.[33] Airports in Changi, Singapore (fig. 120), Edmonton, Canada, and Christchurch, New Zealand, also feature living walls, as does the Etsy headquarters in Dumbo, Brooklyn, where 11,000 indoor plants were installed in 2016 by Greenery NYC.[34] The dramatic installation ensures that every employee can see some of the greenery, which benefits from a state-of-the-art rainwater-harvesting and irrigation system.

Two decades into the twenty-first century, the popular enthusiasm for nature inside continues unabated. It appeals to human beings in very complex ways, some of them probably too deeply embedded in the human psyche for us to be able to easily explain them. Nature inside's life-affirming qualities undoubtedly act as a form of compensation for some of the less life-enhancing aspects of contemporary life, for example, the loss of rural life for large sections of the population, an over-dependence on technology, social isolation, and environmental problems. It also provides us with nurturing opportunities and even a form of friendship. It eases tensions, both psychological and spatial.

While these functions of nature inside have remained constant throughout the modern era, it is also clear that indoor plants and flowers have been able to change their symbolic meanings to suit the dominant agendas of the day. While in the Victorian era they took on a strong religious or spiritual dimension, within modernism their spatial properties were exploited to the full. Although by the 1960s they had become aligned with a scientific agenda, by the following decade their affective powers had come to be fully appreciated. More recently, they have come to represent all the issues being addressed by environmentalists.

Whatever functions and meanings nature inside takes on in the future, it is clear that our desire to bring plants and flowers into our indoor spaces is not going away in the short term. Whether in the home, the workplace, the shopping mall, or in the spaces that operate across the private–public divide, by inviting greenery and flowers inside we feel that we are engaging with nature and helping to re-establish a more balanced relationship with it.

Notes

INTRODUCTION
1. Lind 2004, p. 3.
2. Simons and Ruthven 1995, p. 17.
3. The terms 'modern', 'modernity', 'modernism', 'modernist', 'late modernism' and 'late modernist' can be very confusing. In this book, 'modern' is used as the opposite of what was in the past, and 'modernity' as the condition of the modern era. 'Modernism' refers to a cultural response to modernity that occurred in the inter-war years of the twentieth century in the West, and 'modernist' refers to a protagonist of modernism. 'Late modernism' denotes a later version of modernism, which occurred, especially in the commercial context in the USA, in the years after 1945; a 'late modernist' is one of its protagonists.
4. Baudrillard 1982, quoted in Heynen 1999, p. 29.
5. Plumwood 1993, p. 23.
6. The theory of the 'separate spheres' suggests that, with the arrival of urbanisation and the defined working day in the mid-nineteenth century, a middle-class gender split occurred between men, who went out to the workplace, and women, who stayed at home to mind the children and the house.
7. See Latour 2004. Latour used the word 'actant' in that context, a term which, in 2010, Jane Bennett defined as 'a source of action that can either be human or nonhuman; it is that which has efficacy, can *do* things, has sufficient coherence to make a difference, produce effects, alter the course of events'.
8. Hitchings 2004, p. 110.
9. Plumwood 1993, p. 2. See also Brown and Jordanova 1995.
10. See R. Williams 1983.

CHAPTER 1
TAMING THE JUNGLE
1. Schiebinger 2004, p. 3.
2. Ibid., p. 28.
3. Wulf 2009, p. 207.
4. Ibid., p. 5.
5. Ibid., p. 19.
6. Schiebinger 2004, p. 55.
7. Ibid., p. 11.
8. Lafuente and Valverde 2007, p. 134.
9. Plumwood 1993, p. 158.
10. Schiebinger 2004, pp. 7, 198.
11. See Hunt 1992.
12. Wulf 2009, p. 142.
13. Woods and Warren 1990, p. 55.
14. Schiebinger 2004, p. 58.
15. Ibid., p. 11.
16. Hix 1974, p. 120.
17. On Berlin's botanical gardens, see ibid., p. 168; Koppelkamm 1981, p. 33.
18. Schiebinger 2004, p. 22.
19. Clark 1973, p. 31.
20. Hix 1974, p. 9.
21. Schiebinger 2004, p. 81.
22. Lecture given by Edward Diestelkamp at the study day 'A World under Glass', held at the RHS Lindley Library, London, on 4 May 2018.
23. Ibid.
24. John Loudon 1817, p. 27.
25. Hunt 1992, p. 157.
26. Hibberd 1895, p. 133.
27. See Archiseek n.d.
28. See Hughes 2012.
29. Repton 1816, p. 84.
30. Ibid., p. 87.
31. John Loudon, 2015 [1824], p. 81.
32. H. T. Williams 1872, pp. 77, 81.
33. Wulf 2009, p. 237.
34. Martin 1988, p. 133.
35. Ibid.
36. H. T. Williams 1872, p. 77.

CHAPTER 2
THE JUNGLE IN THE PARLOUR
1. Hibberd 1987 [1856], p. 4.
2. See Dash 2001.
3. See Knowles 2014.
4. See Cohen 2006; Logan 2001; and Nieswander 2008.
5. H. T. Williams 1872, p. 57.
6. Hibberd 1987 [1856], p. 2.
7. H. T. Williams 1872, p. 7.
8. Ibid., p. 8.
9. Sutcliffe 1898, p. 333.
10. Hibberd 1987 [1856], p. 20.
11. K. Williams 2007, p. 6.

12. Ibid., p. 23.
13. H. T. Williams 1872, p. 7.
14. Whittingham 2009, p. 5.
15. See Allen 1969; and Whittingham 2009.
16. Jane C. Loudon's *Instructions in Gardening for Ladies* was published in 1840. It was followed by *The Ladies' Flower Garden* series (1840–48), *Botany for Ladies* (1842), and *The Amateur Gardener's Calendar* (1847).
17. *Instructions in Gardening for Ladies* sold 1,350 copies on its day of publication.
18. See Leavitt 2002.
19. Advice writers on the subject include Hibberd 1987 [1856]; Randolph 1861; Maling 1862a; March 1862; Sprague 1876; H. T. Williams 1872; Burbidge 1874; Hassard 1875; and Mollinson 1877.
20. See Lees-Maffei 2001.
21. British journals of the period included: *The Floral World and Garden Guide* (1825–90); *Gardener's Magazine*, edited by John Loudon (1826–43); *Horticultural Journal* (1833–6), which published the results of floral competitions and which became *Gardener's Gazette* (1836–47); *Gardener's Chronicle* (1841 onwards); *Cottage Garden* (1848–61), which in 1861 became *Journal of Horticulture*; *The Garden*, in which William Robinson wrote a weekly called 'The Garden in the House'; *Queen*; *The Lady*; and *Illustrated London News*. In the USA *Ladies Home Journal* grew out of horticulture or agriculture publications and set about trying to reach a female audience. Annie C. Brown's 'Ferns and Houseplants' (4 May 1887, p. 12) and Eben E. Rexford's 'Houseplants' (1 April 1884, p. 8) were both published in it.
22. Hibberd 1987 [1856], p. 3.
23. Shirley Hibberd also wrote *The Fern Garden* (1869), *New and Rare Beautiful Leaved Plants* (1870), *The Amateur's Flower Garden* (1871) and *The Ivy* (1872). From 1858 he edited *The Floral World*.
24. Hibberd 1987 [1856], pp. xii, 2.
25. Ibid., pp. 3–4.
26. Ibid., p. 177.
27. H. T. Williams 1872, p. 160.
28. Ibid., p. 163.
29. Ibid., p. 247.
30. The Royal Horticultural Society was founded on 7 March 1804, by Sir Joseph Banks and John Wedgwood, as the Horticultural Society of London. It was based in South Kensington. In 1821 an experimental garden was established in Chiswick, and the Society began an extensive series of plant-collecting expeditions which covered the globe and introduced a wealth of plants to Britain. In 1861 the Society was renamed the Royal Horticultural Society after Prince Albert, its then president, arranged a new charter. A new garden in Kensington was also secured, which remained its headquarters until 1888. In 1862 the March daughters followed their father's success of the previous year by winning first prize for 'Drawing Room Decoration', with flowers in baskets with glass handles.
31. See Perkins 1877.
32. Other key texts on the subject by the same author include *Indoor Plants and How to Grow Them, for the Drawing Room, Balcony and Greenhouse*, with three editions published in 1861, 1862 and 1863; and *A Handbook for Ladies: On In-Door Plants, Flowers for Ornament, and Song Birds* of 1867.
33. Maling 1862b, p. 4.
34. Mollinson 1877, p. 161.
35. Ibid., p. i.
36. See Conder 1891.
37. Warner 1917, p. 81.
38. Martin 1988, p. 171.

CHAPTER 3
THE PARLOUR OUTSIDE
1. Relf 1992, p. 11.
2. See Hahn 2010.
3. Pimlott 2016, p. 17.
4. See Downing, 2009.
5. Munro 1971, p. 6.
6. Koppelkamm 1981, p. 29.
7. Woods and Warren 1990, p. 128.
8. Ibid.
9. Ibid.
10. Koppelkamm 1981, p. 31. Quotation from Woods and Warren 1990, p. 128.
11. Koppelkamm 1981, p. 46.
12. Hix 1974, p. 122.
13. See Beaver 2001.
14. See Valen 2016.
15. See Nichols and Turner 2017.
16. See Munro 1971.
17. Phillips 1875, p. 6.
18. *The Era* 1876, p. 4.
19. See Shaw 1890.
20. Ibid., pp. 11, 13.
21. Examples included the tanks in Regent's Park Gardens, added in 1853; the fish tanks in the Sydenham Crystal Palace of 1871; and the Brighton Aquarium, opened a year later.
22. Munro 1971, p. 10.
23. See Shaw 1890.
24. Munro 1971, p. 10.
25. See Walton 2000.
26. See Potter 1999.
27. Ibid., p. 6.
28. See Barker n.d.
29. The Langham Hotel, designed by John Giles, was built between 1863 and 1865 at a cost of £300,000.
30. After nearly five years of building, The Savoy opened on 6 August 1889. Built by Richard D'Oyly Carte on land adjacent to his Savoy Theatre, the Savoy Hotel offered accommodation for the many tourists, especially Americans, who travelled to London to see the Savoy Operas.
31. The Carlton, which in its early days was one of London's most fashionable hotels, was originally run by the Swiss hotelier César Ritz with Auguste Escoffier as the head chef. Previously the pair had managed The Savoy, from which The Ritz poached a certain amount of custom.
32. Denby 2002, p. 148.
33. Ibid., p. 150.
34. Ibid., p. 156.
35. Rappaport 2000, p. 10.

CHAPTER 4
NATURAL MODERNISM
1. Lancaster 1939, p. 76.
2. Todd and Mortimer 1929, p. 2.
3. Overy 2007, p. 180.
4. Wilk 2006, p. 325.
5. Ibid., p. 323.
6. See Overy 2007.
7. Ibid., p. 124.
8. See Rosner 2008.
9. See Bonaiti 2004.
10. Overy 2007, p. 325.
11. Isenstadt 2014, p. 300.
12. Overy 2007, p. 135.

13. The 'liner' look became a feature of the popular idiom known as Art Deco, which took modernism to a mass audience in the 1930s. In the USA the equivalent style was known as 'streamform'.
14. See Overy 2007.
15. Ibid., p. 125.
16. Ibid., p. 105.
17. See Abernathy 1988.
18. See Nerdinger 2019.
19. Dümpelmann and Beardsley 2015, p. 97.
20. See Dina 2008.
21. Connors 1984, p. 1.
22. Gibson 2017, p. 2.
23. Pallasmaa 2003, p. 57.
24. Suominen-Kokkonen 2007, p. 128.
25. Wilk 1981, p. 63.
26. Charles Sheeler's 1931 painting *Cactus* is in the Philadelphia Museum of Art.
27. H. Hoffmann 1930, pp. 25, 62 and 173.
28. See Da Costa Meyer 2016.
29. Lamplugh 1929, p. 18.
30. Ibid., p. 21.
31. Spry 1952, p. 12.
32. *The Home of Today* 1938, p. 520.
33. Rockwell and Grayson 1947, p. 14.

CHAPTER 5
THE GARDEN INSIDE
1. Meeker 1988, p. 31.
2. See Overy 2007.
3. See Stiller 1999; Hammer-Tugendhat and Tegethoff 2000; Winter 2000; Neumann 2007; Polan 2008; Neuhoff 2009; Nierhaus 2009; Cernouskova 2012; and Schulze and Windhorst 2012.
4. Hammer-Tugendhat and Tegethoff 2000, p. 5.
5. See Riezler 1931.
6. See ibid.; Stiller 1999; Winter 2000; Neuhoff 2009; and Nierhaus 2009.
7. Hammer-Tugendhat and Tegethoff 2000, p. 1.
8. Ibid., p. 55.
9. See Günther 1988.
10. See Lange 2006.
11. See Eggler-Gerozissis 2009.
12. Ibid.
13. Hammer-Tugendhat and Tegethoff 2000, p. 19.
14. Ibid., p. 16.
15. McQuaid 1996, p. 22.
16. Ibid., p. 22.
17. Hammer-Tugendhat and Tegethoff 2000, p. 8.
18. Ibid., p. 36.
19. See Villa Tugendhat n.d.
20. On the spiritual quality of the villa, see Hammer-Tugendhat and Tegethoff 2000, p. 109.

CHAPTER 6
LIVING IN THE GARDEN
1. Flanders and Swann 1956.
2. Kaplan 1993, p. 87.
3. Ibid., p. 32.
4. Ibid.
5. Sadar 2016, p. 182.
6. Neutra 1935, p. 15.
7. Leet 2004, p. 19.
8. See Marx 1964.
9. Leet 2004, p. 76.
10. Ibid., p. 108.
11. Ibid., pp. 118–19.
12. See Leet 2004.
13. See Serraino 2000; and Alexander 2011.
14. Leet 2004, p. 147.
15. A. Friedman 2010, p. 87.
16. Shulman 1977, pp. 157, 217.
17. See McCoy 1977; and Smith 2006.
18. Ince and Johnson 2015, p. 24.
19. Kirkham 1998, p. 143.
20. Ince and Johnson 2015, p. 103.
21. Eckbo 1978, p. 9. The first edition was published in 1956.
22. See Woo 2010.
23. See Online Archive of California n.d.
24. Kaplan 1993, p. 150.
25. Ibid., p. 191.
26. *Arts & Architecture*, November 1949, p. 10.
27. Eckbo 1978, p. 43.
28. *Arts & Architecture*, December 1949, p. 33.

CHAPTER 7
NATURAL LATE MODERNISM
1. Manaker 1981, p. 2.
2. See Zunz 1990.
3. Anderson 1955, p. 3.
4. Marx 1964, p. 5.
5. See Stephens 1978.
6. For more detail about the Four Seasons see B. H. Friedman 2009; Mariani and von Binder 1999; Filler 2009; and Lanks and Makovsky 2009.
7. See Lambert 2005.
8. See Krebs 1987.
9. Mariani and von Binder 1999, p. 26.
10. Ibid., p. 27.
11. See Linn 2007.
12. Mariani and von Binder 1999, p. xv.
13. Ibid., p. 29.
14. For details of the life of Everett L. Conklin see his obituary, *New York Times* 1985.
15. See Pelkonen 2011.
16. For the story of the American mall see Kowinski 1985; and Maitland 1990.
17. See Gruen and Baldauf 2017.
18. See R. Wilson 2004; and Halkias 2015.
19. See Wynn 2017.
20. See Halprin 2011; and Helphand 2017.
21. See Gibbins 2015.
22. Gissen 2014, p. 69.
23. Ibid., p. 6.
24. See McQuade 1964; Huxtable 1967; Faust 1968; and Gissen 2014 (chapter 2).
25. See Kiley and Amidon 1999.
26. See Hildebrand and Dillon 1998.
27. Huxtable 1967, p. 10.
28. See Conklin 1978.
29. Ibid., p. 73.
30. Ibid., p. 74.
31. Gissen 2014, p. 68.
32. Gissen 2014, p. 83.

CHAPTER 8
THE LIVING ROOM IN THE CITY
1. Bednar 1986, p. 67.
2. See Riani and Goldberger 1990; Goldberger and Craig 2009; and Rice 2016.
3. Portman and Barnett 1976, p. 60.
4. Portman et al. 1997, p. 16.
5. Ibid., p. 17.
6. Ibid., p. 19.
7. Ibid., p. 21.
8. Barnett 1966, p. 139.
9. Bednar 1986, pp. 15–18.
10. Ibid., p. 17.
11. Portman and Barnett 1976, p. 71.
12. *Engineering News Record* 1965, p. 26.
13. Ibid.
14. Goetz 2003, p. 2.
15. Ibid.
16. Ibid.
17. *The Atlanta Journal* 1967, p. 1.

18. See Carter 1998.
19. *The Atlanta Journal* 1967, p. 8.
20. Ibid., p. 2.
21. Ibid., p. 3.
22. Goetz 2003, p. 2.
23. *The Atlanta Journal* 1967, p. 3.
24. Ibid., p. 1.
25. Portman and Barnett 1976, p. 130.
26. Barnett 1966, p. 139.
27. Goetz 2003, p. 2.
28. Quotations from Barnett 1966, p. 139; and Goetz 2003, p. 2.
29. *Interiors*, July 1967, p. 69.
30. Portman and Barnett 1976, p. 96.
31. *The Atlanta Journal* 1967, p. 13.
32. Ibid.
33. Ibid., p. 17.

CHAPTER 9
THE BENEFITS OF NATURE INSIDE

1. Plumwood 1993, p. 69.
2. See Graf 1976.
3. Hammer 1992, p. 12.
4. Ibid.
5. Ibid.
6. See Gaines 1977.
7. Ibid., p. ix.
8. Ibid.
9. Ibid., p. x.
10. Ibid., p. xiv.
11. Ibid., p. xv.
12. Manaker 1981, p. ix.
13. Ibid., pp. 1–2.
14. See Furuta 1983.
15. Ibid.
16. Fromm 1964, p. 38.
17. Conklin 1972.
18. Conklin 1978, p. 74.
19. Ibid.
20. E. Wilson 1984, p. 3.
21. Ibid., p. 10.
22. Ibid., p. 139.
23. Ibid., p. 113.
24. Ibid., p. 25.
25. Ibid., p. 118.
26. Kellert and Wilson 1993, p. 21.
27. Ibid., p. 20.
28. Ibid., p. 14.
29. Relf 1992, p. 95.
30. Ulrich 1979, p. 19.
31. They included *Scandinavian Journal of Psychology* (established in 1960), *Landscape Research* (established in 1968), *Environment and Behavior* (established in 1969), *Landscape and Urban Planning* (established in 1974), *Journal of Environmental Psychology* (established in 1980), *Journal of Housing for the Elderly* (established in 1983), *Indoor and Built Environment* (established in 1992), *Journal of Home and Consumer Horticulture* (established in 1993), *The Journal of Horticultural Science and Biotechnology* (established in 1998, formerly known as the *Journal of Horticultural Science* [1948–97]), *North American Journal of Psychology* (established in 1999), *Evolutionary Psychology* (established in 2003) and *International Journal of Environmental Research and Public Health* (established in 2004), the large number being a measure of the academic and professional interest in the field.
32. Saval 2014, pp. 201–2.
33. Ibid., p. 204.
34. Conklin 1978, p. 73.
35. Kaplan 1993, p. 196.
36. Ibid., p. 199.
37. Fjeld et al. 1998.
38. Evenson et al. 2015, p. 289.
39. Talbott et al. 1976, pp. 365–6.
40. Ulrich 1984, p. 420.
41. Park and Mattson 2009, p. 105.
42. See Holmes 2013.
43. See Relf 1992.
44. See TLCstaff 2011.
45. Ibid.
46. See Tifferet and Vilnai-Yavetz 2017.
47. Ibid., p. 844.
48. Wolverton, Johnson and Bounds 1989, p. 2.
49. Ibid., p. 18.
50. See Wolverton 1996.

CHAPTER 10
GREENING THE INTERIOR

1. Lewis 1992, p. 11.
2. Carson 1962, p. 11.
3. Ibid., p. 15.
4. Ibid., p. 77.
5. Morton 2007, p. 9.
6. See Veder 2007.
7. Mollinson 1877, p. 2.
8. See Hammer 1992, p. 62.
9. Hibberd 1987 [1856], p. 2.
10. Morton 2007, p. 14.
11. See Wright 1985.
12. Bonar 1977, p. 12.
13. Ibid., p. 14.
14. Ibid., p. 15.
15. Ibid., p. 42.
16. Rees 1995, p. 6.
17. Gilliatt 1986, p. 84.
18. Bruun and Kalland 1995, p. 10.
19. Ibid., p. 11.
20. See Turner 2009.
21. Dornob Staff n.d.
22. See Archello n.d.
23. See Gibson 2018.
24. See Howarth 2017.
25. Gottlieb 1984, p. 25.
26. See Reyes 2017.
27. English 2011, p. 1.
28. Ibid.
29. Foster + Partners n.d.
30. See Sky Garden n.d.
31. See Moore 2015.
32. See Biotecture 2013.
33. Cosgrove 2016, p. 1.
34. See Silva 2017.

Bibliography

Abernathy, Ann, 1988, *The Oak Park Home and Studio of Frank Lloyd Wright* (Chicago: Frank Lloyd Wright Preservation Trust).

Adachi, M., C. L. E. Rohde and A. D. Kendle, 2000, 'Effects of Floral and Foliage Displays on Human Emotions', *HortTechnology* 10, no. 1, pp. 59–63.

Alexander, Christopher James, and Wim de Wit, 2011, *Julius Shulman's Los Angeles* (Los Angeles: J. Paul Getty Trust).

Allen, David Elliston, 1969, *The Victorian Fern Craze: A History of Pteridomania* (London: Hutchinson).

Anderson, Harvey, 1955, 'Residential to Non-Residential', *Interior Design* 26 (April), p. 3.

Anker, Peder, 2010, *From Bauhaus to Ecohouse: A History of Ecological Design* (Baton Rouge: Louisiana State University Press).

Archello, n.d., 'Elok House', Archello (website), https://archello.com/project/elok-house (accessed 18 March 2020).

Archiseek, n.d., '1754 – Bellevue House, Delgany, Co. Wicklow', Archiseek (website), http://archiseek.com/2015/1754-bellevue-house-delgany-co-wicklow/ (accessed 18 March 2020).

The Atlanta Journal, 1967, 'The Hyatt Regency Hotel', 25 June 1967, pp. 1–17.

Augé, Marc, 1995, *Non-Places: An Introduction to Supermodernity* (London and New York: Verso).

Balmori, Diana, and Joel Sanders, 2011, *Groundwork: Between Landscape and Architecture* (New York: Monacelli Press).

Barker, Darren, n.d., 'The History of the Winter Gardens in Great Yarmouth', Brickwork (website), https://brick-work.org/the-history-of-the-winter-gardens-in-great-yarmouth/ (accessed 18 March 2020).

Barnett, Jonathan, 1966, 'John Portman: Atlanta's One-Man Urban Renewal Programme' *Architectural Record* 32, no. 1 (January), pp. 136-41.

Baudrillard, Jean, 1982, 'Modernité', in *La Modernité ou l'esprit du temps*, Catalogue of the 12th Biennale de Paris, Section Architecture (Paris: L'Equerre), pp. 27–34.

Beaver, Patrick, 2001, *The Crystal Palace: A Portrait of a Victorian Enterprise* (Chichester: Phillimore).

Bednar, Michael J., 1986, *The New Atrium* (New York: McGraw Hill).

Beecher, Catharine, and Harriet Beecher Stowe, 1996 [1869], *The American Woman's Home, or, Principles of Domestic Science* (New Brunswick, NJ: Rutgers University Press).

Bennett, Jane, 2010, *Vibrant Matter: A Political Ecology of Things* (Durham, NC, and London: Duke University Press).

Bennett, Jennifer, 1991, *Lilies of the Heart: The Historical Relationship between Women and Plants* (Camden East, Ont.: Camden House Publishing).

Benton, Tim, 2006a, 'Modernism and Nature', in *Modernism: Designing a New World*, ed. Christopher Wilk (London: V&A Publications), pp. 311–40.

—, 2006b, *The Modernist Home* (London: V&A Publications).

Bernhardt, Peter, 1999, *The Rose's Kiss: A Natural History of Flowers* (Washington, DC: Island Press).

Berrall, Julia S., 1953, *A History of Flower Arrangement* (London: Thames and Hudson).

Biggs, Caroline, 2018, 'Plant-Loving Millennials at Home and Work', *New York Times* (online), 9 March 2018, www.nytimes.com/2018/03/09/realestate/plant-loving-millenials-at-home-and-at-work.html, (accessed 18 March 2020).

Biotecture, 2013, 'Office Interior, Wilson Street', Biotecture (website), www.biotecture.uk.com/portfolio/office-interior-wilson-street (accessed 18 March 2020).

Blacker, Mary Rose, 2000, *Flora Domestica: A History of British Flower Arranging, 1500–1930* (London: National Trust).

Block, India, 2018, 'Stefano Boeri Designs First Tree-Covered Social-Housing Project', *Dezeen*, 16 January 2018, www.dezeen.com/2018/01/16/stefano-boeri-trudo-vertical-forest-social-housing-tower-eindhoven-plants-trees/ (accessed 18 March 2020).

Bonaiti, Maria, 2004, *Le Corbusier et la Nature* (Paris: Editions de la Villette).

Bonar, Ann, 1977, *The St Michael Guide to Houseplants* (London: Hennerwood Publications).

Bringslimark, Tina, Terry Hartig and Grete G. Patil, 2009, 'The Psychological Benefits of Indoor Plants: A Critical Review of the Experimental Literature', *Journal of Environmental Psychology* 29, no. 4, pp. 422–33.

Brown, Joseph, 2007, 'A Nostalgic Look at the Americana Hotel', *South Beach Magazine* (online), 11 July 2007, www.southbeachmagazine.com/a-nostalgic-look-at-the-americana-hotel/ (accessed 18 March 2020).

Brown, Julia Prewitt, 2008, *The Bourgeois Interior: How the Middle-Class Imagines Itself in Literature and Film* (Charlottesville and London: University of Virginia Press).

Brown, Penelope, and Ludmilla Jordanova, 1995, 'Oppressive Dichotomies: The Nature/Culture Debate', in *A Cultural Studies Reader: History, Theory, Practice*, ed. Jessica Munns, Gita Rajan and Roger Bromley (London and New York: Routledge), pp. 509–18.

Bruun, Ole, and Arne Kalland (eds), 1995, *Asian Perceptions of Nature: A Critical Approach* (London: Routledge).

Burbidge, Frederick William, 1874, *Domestic Floriculture: Window Gardening and Floral Decorations, being Practical Directions for the Propagation, Culture and Arrangement of Plants and Flowers as Domestic Ornaments* (Edinburgh and London: William Blackwood and Sons).

Cabot Smith, Ronald, 2013, *Interior Plantscaping with Large Houseplants* (Fargo: North Dakota State University Extension Service).

Canty, Donald, 1974, 'Office Landscaping: Idea to Industry in Ten Years', *AIA Journal* (October), pp. 19–24.

Carroll, Khadija von Zinnenburg (ed.), 2018, *Botanical Drift: Protagonists of the Invasive Herbarium* (Berlin: Sternberg Press).

Carson, Rachel, 1962, *Silent Spring* (Harmondsworth: Penguin Books).

Carter, Brian, 1998, *Johnson Wax Administration Building* (London: Phaidon).

Cernouskova, Dagmar, 2012, 'Brno's Villa Tugendhat: Eight Decades of a Modern Residence', *Journal of Architectural and Town-Planning Theory* 36, nos 1–2, pp. 24–51.

Clark, Frank, 1973, 'Nineteenth-Century Public Parks from 1830', *Garden History* 1, no. 3, pp. 31–41.

Clough, Patricia Tiniceto, and Jean Halley (eds), 2007, *The Affective Turn: Theorizing the Social* (Durham, NC: Duke University Press).

Coats, A. M., 1969, *The Plant Hunters; Being a History of the Horticultural Pioneers, Their Quests, and Their Discoveries from the Renaissance to the Twentieth Century* (New York: McGraw Hill).

Cohen, Deborah, 2006, *Household Gods: The British and Their Possessions* (New Haven and London: Yale University Press).

Collins, Barbara L., 2002, *Professional Interior Plantscaping* (Champaign, IL: Stipes).

Conder, Josiah, 1891, *The Flowers of Japan and the Art of Floral Arrangement* (Tokyo: Hakubunsha).

Conklin, Everett L., 1972, 'Man and Plants: A Primal Association', *American Nurseryman Magazine* 136, no. 9, pp. 46–9.

—, 1974, 'Interior Plantings Bring Nature Indoors', *American Nurseryman Magazine* 139, no. 2, pp. 12–13.

—, 1978, 'Interior Landscaping', *Journal of Arboriculture* 4, no. 4, pp. 73–9.

Connors, Joseph, 1984, *The Robie House of Frank Lloyd Wright* (Chicago: University of Chicago Press).

Conran, Terence, 1990, *Conran's Decorating with Plants* (London: Smithmark).

Cook, Jenny, 1993, 'Bringing the Outside In: Women and the Transformation of the Middle-Class Maritime Canadian Interior, 1830–1860', *Material Culture Review* 38 (Fall), pp. 36–49.

Corner, E. J. H., 1966, *The Natural History of Palms* (Berkeley and Los Angeles: University of California Press).

Cosgrove, Sarah, 2016, 'Heathrow Airport Introduces Green Walls to Relax Stressed Passengers', *Horticulture Week*, 10 October 2016, www.hortweek.com/heathrow-airport-introduces-green-walls-relax-passengers/landscape/article/1411661 (accessed 18 March 2020).

Crook, Lizzie, 2018, 'Foster + Partners Use Trees as Partitions Inside Thailand's First Apple Store', *Dezeen*, 13 December 2018, www.dezeen.com/2018/12/13/foster-partner-thailand-apple-iconsiam-store/ (accessed 18 March 2020).

Da Costa Meyer, Esther, 2016, *Pierre Chareau: Modern Architecture and Design* (New Haven and London: Yale University Press).

Darwin, Charles, 2003 [1859], *On the Origin of Species* (London: Signet).

Dash, Mike, 2001, *Tulipomania: The Story of the World's Most Coveted Flower and the Extraordinary Passions It Aroused* (New York: Broadway Books).

Davies, Jennifer, 1991, *The Victorian Flower Garden* (London: BBC Books).

DelPrince, James M., 2012, *Interior Plantscaping: Principles and Practices* (New York: Delmar Cengage Learning).

Demos, T. J., 2016, *Decolonizing Nature: Contemporary Art and the Politics of Ecology* (Berlin: Sternberg Press).

Denby, Elaine, 2002, *Grand Hotels: Reality and Illusion* (London: Reaktion Books).

Dennis, Lori, 2010, *Green Interior Design* (New York: Allworth Press).

Descola, Philippe, 2013, *The Ecology of Others* (Chicago: Prickly Paradigm Press).

Dina, Lucia Borromeo, (ed.), 2008, *Villa Necchi Campiglio a Milano* (Milan: Skira).

Domec, Laurent, 2008, *La Grande Aventure des plantes d'intérieur: Histoire et symbolisme des origines à nos jours* (Paris: Éditions Alternatives).

Dornob Staff, n.d., 'Amazing Home Atrium & Multi-Level Interior Garden Design', Dornob (website), https://dornob.com/amazing-home-atrium-multi-level-interior-garden-design/ (accessed 18 March 2020).

Downing, Sarah Jane, 2009, *The English Pleasure Garden, 1660–1860* (Oxford: Shire Books).

Duffey, Frank, 1969, *Office Landscaping: A New Approach to Office Planning* (London: Anbar).

Dümpelmann, Sonja, and John Beardsley (eds), 2015, *Women, Modernity and Landscape Architecture* (London: Routledge).

Dunham, Donald, 2012, 'Architecture without Nature', in *Earth Perfect? Nature, Utopia and the Garden*, ed. Annette Giesecke and Naomi Jacobs (London: Black Dog), pp. 136–55.

Eckardt, Marianne Horney, 1992, 'Fromm's Concept of Biophilia', *Journal of the American Academy of Psychoanalysis* 20, no. 2, pp. 233–40.

Eckbo, Garrett, 1978, *The Art of Home Landscaping*, revised and enlarged edition (New York: McGraw Hill).

Eckstein, Hans, 1941, *Die schöne Wohnung: Wohnräume der Gegenwart in 175 Abbildungen* (Munich: F. Bruckmann Verlag).

Eggler-Gerozissis, Marianne, 2009, 'Divide and Conquer: Ludwig Mies van der Rohe and Lilly Reich's Fabric Partitions at the Tugendhat House', *Studies in the Decorative Arts* 16, no. 2, pp. 66–90.

Engineering News Record, 1965, 'The Hyatt Regency Hotel', 29 July 1965, pp. 23–9.

English, Rebecca, 2011, 'How Westminster Abbey is being Turned into a £50k Fairytale Forest for the Royal Wedding', *Daily Mail* (online), 28 April 2011, www.dailymail.co.uk/news/article-1381134/Royal-wedding-2011-Westminster-Abbey-turned-50k-fairytale-forest.html (accessed 18 March 2020).

The Era, 1876, 'Opening of the Royal Aquarium', 30 January 1876, p. 4.

—, 1883, 'The Royal Aquarium', 6 January 1883, p. 11.

Evans, M. E., 1992, 'Plants and People: A Case-Study in the Hotel Industry', in *The Role of Horticulture in Human Well-Being and Social Development*, ed. Diane Relf (Portland, OR: Timber Press), pp. 220–22.

Evenson, Katinka, et al., 2015, 'Restorative Elements at the Computer Workstation: A Comparison of Live Plants and Inanimate Objects With and Without Window View', *Environment and Behavior* 47, no. 3, pp. 288–303.

Falkenberg, Haike, 2011, *Interior Gardens: Designing and Constructing Green Spaces in Private and Public Buildings* (Basel: Birkhauser Verlag).

Faust, Joan Lee, 1968, 'The Outside-In Building', *New York Times*, 17 March 1968, sec. Arts, p. 39.

Fediw, Kathy, 2015, *The Manual of Interior Plantscaping: A Guide to Design, Installation, and Maintenance* (Portland, OR: Timber Press).

Felski, Rita, 1995, *The Gender of Modernity* (Cambridge, MA: Harvard University Press).

Filler, Martin, 2009, 'Serving Up a Heady Cocktail of Gravitas and Glamour', *Architectural Record* 194, no. 9, pp. 66–74.

Fjeld, Tove, et al., 1998, 'The Effect of Indoor Foliage Plants on Health and Discomfort Symptoms among Office Workers', *Indoor and Built Environment* 7, no. 4, pp. 204–9.

Flanders, Michael, and Donald Swann, 1956, 'Design for Living', *At the Drop of a Hat* (vinyl LP), Parlophone PMCO 1033.

Foster + Partners, n.d., 'Apple Orchard Road, Singapore', Foster + Partners (website), www.fosterandpartners.com/projects/apple-orchard-road-singapore/ (accessed 18 March 2020).

Francis, Mark, and Randolph T. Hester Jr. (eds), 1990, *The Meaning of Gardens: Idea, Place, and Action* (Cambridge, MA: MIT Press).

Friedman, Alice T., 2010, *American Glamour and the Evolution of Modern Architecture* (New Haven and London: Yale University Press).

Friedman, B. H., 2009, 'The Most Expensive Restaurant Ever Built', *The Evergreen Review* 120, October 2009, https://evergreenreview.com/read/the-most-expensive-restaurant-ever-built/ (accessed 18 March 2020).

Fritsch, Matthias, Philippe Lynes and David Wood (eds), 2018, *Eco-Deconstruction: Derrida and Environmental Philosophy* (New York: Fordham University Press).

Fromm, Erich, 1964, *The Heart of Man: Its Genius for Good and Evil* (New York, Evanston and London: Harper & Row).

Furuta, Tok, 1983, *Interior Landscaping* (Reston, VA: Reston Publishing).

Fry, Carolyn, 2009, *The Plant Hunters* (London: Andre Deutsch).

Gaines, Richard L., 1977, *Interior Plantscaping: Building Design for Interior Foliage Plants* (New York: Architectural Record Books).

Gere, Charlotte, 2010, *Artistic Circles: Design and Decoration in the Aesthetic Movement* (London: V&A Publishing).

Gibbins, Kristen, 2015, 'The Story behind the Iconic NorthPark Planters', NorthPark Center (website), 14 January 2015, http://northparkcenter.com/posts/northpark-the-story-behind-the-iconic-northpark-planters (accessed 18 March 2020).

Gibbons, Martin, 2000, *Identifying Palms: The New Compact Study Guide and Identifier* (London: Apple).

Gibson, Eleanor, 2017, 'Frank Lloyd Wright's Hollyhock House is an Early Example of Mayan Revival Architecture', *Dezeen*, 6 June 2017, www.dezeen.com/2017/06/06/hollyhock-house-frank-lloyd-wright-los-angeles-california-mayan-revival-architecture/ (accessed 18 March 2020).

—, 2018, 'David Guerra Wraps Brazilian House around Courtyard Filled with Tropical Plants', *Dezeen*, 3 February 2018, www.dezeen.com/2018/02/03/valley-house-ii-david-guerra-nova-lima-minas-gerais-brazil-tropical-courtyard-garden/ (accessed 18 March 2020).

Gilliatt, Mary, 1986, *English Country Style* (London: Macdonald Orbis).

Gissen, David, 2014, *Manhattan Atmospheres: Architecture, the Interior Environment, and Urban Crisis* (Minneapolis and London: University of Minnesota Press).

Goetz, Alisa, 2003, *Up, Down and Across: Elevators, Escalators and Moving Sidewalks* (Washington, DC: National Building Museum).

Goldberger, Paul, and Robert M. Craig, 2009, *John Portman: Art and Architecture* (Atlanta: High Museum of Art / University of Georgia Press).

Goody, Jack, 1993, *The Culture of Flowers* (Cambridge: Cambridge University Press).

Gottlieb, Martin, 1984, 'Palms and a Pavilion for Battery Park City', *New York Times*, 11 August 1984, sec. 1, p. 25.

Graf, Alfred Byrd, 1976, *Exotic House Plants Illustrated*, 10th edition (East Rutherford, NJ: Roehrs).

Gribbin, Mary, and John Gribbin, 2008, *Flower Hunters* (Oxford: Oxford University Press).

Grier, Katherine C., 1997, *Culture and Comfort: Parlor Making and Middle-Class Identity, 1850–1930* (Washington, DC and London: Smithsonian Institution Press).

Griffiths, Alyn, 2018, 'Le House Fills An'garden Café with Plants to Create an Oasis in Bustling Hanoi', *Dezeen*, 7 April 2018, www.dezeen.com/2018/04/07/le-house-angarden-cafe-planting-architecture-hanoi-vietnam/ (accessed 18 March 2020).

Gruen, Victor, and Anette Baldauf, 2017, *Shopping Town: Designing the City in Suburban America* (Minneapolis: University of Minnesota Press).

Günther, Sonja, 1988, *Lilly Reich, 1885–1947: Innenarchitektin, Designerin, Ausstellungsgestalterin* (Stuttgart: Deutsche Verlags-Anstalt).

Hadfield, Miles, 1997, *The English Landscape Garden* (Aylesbury: Shire Publications).

Hagan, Susannah, 2001, *Taking Shape: A New Contract between Architecture and Nature* (Oxford: Architectural Press).

Hahn, Hazel H., 2010, *Scenes of Parisian Modernity: Culture and Consumption in the Nineteenth Century* (London: Palgrave).

Halkias, Maria, 2015, 'As Other Malls Die Out, NorthPark Turns 50 in Style', *Dallas Morning News* (online), 17 August 2015, www.dallasnews.com/business/retail/2015/08/17/as-other-malls-die-out-northpark-turns-50-in-style/ (accessed 18 March 2020).

Halprin, Lawrence, 2011, *A Life Spent Changing Places* (Philadelphia: University of Pennsylvania Press).

Hammer, Nelson, 1992, *Interior Landscape Design* (New York: McGraw Hill).

—, 1999, *Interior Landscapes: An American Portfolio of Green Environments* (Gloucester, MA: Rockport).

Hammer-Tugendhat, Daniela, and Wolf Tegethoff (eds), 2000, *Ludwig Mies van der Rohe – The Tugendhat House* (Vienna: Springer Verlag).

Hansen, Eric, 2001, *Orchid Fever: A Horticultural Tale of Love, Lust and Lunacy* (London: Methuen).

Haraway, Donna, and Thyrza Nichols Goodeve, 1999, *How Like a Leaf: An Interview with Donna Haraway* (Oxford and New York: Routledge).

Hartl, H., 1929, *Modern Interiors in Colour* (Stuttgart: Julius Hoffmann).

Hassard, Annie, 1875, *Floral Decorations for the Dwelling House: A Practical Guide for the Home Arrangement of Plants and Flowers* (London: Macmillan).

Heath, Oliver, 2018, 'Get Inspired by WELL', Human Spaces (blog), 6 July 2018, https://blog.interface.com/en-uk/get-inspired-well/ (accessed 18 March 2020).

Helphand, Kenneth I., 2017, *Lawrence Halprin* (Athens: University of Georgia Press).

Heynen, Hilde, 1999, *Architecture and Modernity: A Critique* (Cambridge, MA: MIT Press).

Hibberd, Shirley, 1895, *Rustic Adornments for Homes of Taste*, new edition revised by T. W. Sanders (London: W. H. & L. Collingridge).

—, 1987 [1856], *Rustic Adornments for Homes of Taste*, facsimile of the 1856 first edition (London: Century in association with the National Trust).

—, 2012 [1870], *New and Rare Beautiful Leaved Plants* (London: General Books).

—, 2012 [1872], *The Ivy: A Monograph* (London: Facsimile).

—, 2013 [1871], *The Amateur's Flower Garden* (London: Facsimile).

—, 2019 [1869], *The Fern Garden: How to Keep, Make and Enjoy It* (New Delhi: Gyan Books).

Hildebrand, Gary R., and David Dillon, 1998, *The Miller Garden: Icon of Modernism* (Washington, DC: Spacemaker Press).

Hill, Kay, 2018, 'Focus: Biophilia', *Design/Curial*, 27 March 2018, www.designcurial.com/news/focus-biophilia-6094727 (accessed 18 March 2020).

Hillhouse, Lizzie Page, 1897, *House Plants and How to Succeed with Them: A Practical Handbook* (New York: A. T. Delamare).

Hitchings, Russell, 2004, 'At Home with Someone Nonhuman', *Home Cultures* 1, no. 2, pp. 169–86.

—, and V. Jones, 2004, 'Living with Plants and the Exploration of Botanical Encounter in Human Geography Research', *Ethics, Place and Environment* 7, no. 1, pp. 3–19.

Hix, John, 1974, *The Glass House* (Cambridge, MA: MIT Press).

Hoffmann, Donald, 1978, *Frank Lloyd Wright's Fallingwater: The House and Its History* (London: Dover Publications).

—, 1984, *Frank Lloyd Wright's Robie House* (London: Dover Publications).

Hoffmann, Herbert, 1930, *Modern Interiors in Europe and America* (London: Studio).

Holmes, Damian, 2013, 'Crown Sky Garden, Chicago USA: Mikyoung Kim Design', World Landscape Architecture, 9 August 2013, http://worldlandscapearchitect.com/crown-sky-garden-chicago-usa-mikyoung-kim-design/ (accessed 18 March 2020).

The Home of Today: Its Choice, Planning, Equipment and Organisation, 1938 (London: *Daily Express* Publications).

Horsley, Carter B., 1979, 'Some Second Thoughts on Open-Plan Offices', *New York Times*, 18 March 1979, sec. R, p. 1.

Horwood, Catherine, 2007, *Potted History: The Story of Plants in the Home* (London: Frances Lincoln).

—, 2010, *Gardening Women: Their Stories from 1600 to the Present* (London: Virago).

Howarth, Dan, 2017, 'Penda Proposes Toronto Tree Tower Built from Cross-Laminated Timber Modules', *Dezeen*, 2 August 2017, www.dezeen.com/2017/08/02/toronto-tree-tower-penda-cross-laminated-timber-construction/ (accessed 18 March 2020).

Hughes, Tyler, 2012, '"Beacon Hill House": The Newport Cottage of Arthur Curtiss James', The Gilded Age Era (blog), 29 August 2012, http://thegildedageera.blogspot.com/2012/08/beacon-hill-house-newport-cottage-of.html (accessed 18 March 2020).

Hunt, John Dixon, 1992, *Gardens and the Picturesque: Studies in the History of Landscape Architecture* (Cambridge, MA: MIT Press).

Huxtable, Ada Louise, 1967, 'Architecture: Ford Flies High', *New York Times*, 26 November 1967, sec. D, p. 23.

Ince, Catherine, with Lotte Johnson (eds), 2015, *The World of Charles and Ray Eames*, exh. cat. (London: Thames and Hudson / Barbican Art Gallery).

Isenstadt, Sandy, 2014, *The Modern American House: Spaciousness and Middle-Class Identity* (Cambridge: Cambridge University Press).

Jekyll, Gertrude, 1907, *Flower Decoration in the House* (London: Antique Collectors' Club).

—, 2011 [1900], *Home and Garden: Notes and Thoughts, Practical and Critical, of a Worker in Both* (Cambridge: Cambridge University Press).

Johnson, Louisa, 1844, *Every Lady Her Own Flower Gardener* (New Haven: S. Babcock).

Jones, C. F., and Henry T. Williams, 1876, *Household Elegancies: Suggestions in Household Art and Tasteful Home Decorations* (New York: Henry T. Williams).

Josifovic, Igor, and Judith de Graaff, 2016, *Urban Jungle: Living and Styling with Plants* (London: Callwey).

Kahle, Katharine Morrison, 1930, *Modern French Decoration* (New York: G. P. Putnam's Sons).

Kaplan, Rachel, 1993, 'The Role of Nature in the Context of the Workplace', *Landscape and Urban Planning* 26, nos 1–4, pp. 193–201.

—, and Stephen Kaplan, 1989, *The Experience of Nature: A Psychological Perspective* (Cambridge: Cambridge University Press).

Kaplan, Stephen, 1995, 'The Restorative Benefits of Nature: Toward an Integrative Framework', *Journal of Environmental Psychology* 15, no. 3, pp. 169–82.

Kaplan, Wendy (ed.), 2011, *Californian Design, 1930–1965: Living in a Modern Way*, exh. cat. (Los Angeles: LACMA).

Kaufmann-Buhler, Jennifer, 2016, 'Progressive Partitions: The Promises and Problems of the American Open Plan Office', *Design and Culture* 8, no. 2, pp. 205–33.

Kayden, Jerold S., The New York City Department of City Planning, and The Municipal Art Society of New York, 2000, *Privately Owned Public Space: The New York City Experience* (New York: Wiley).

Kellert, Stephen R., and Edward O. Wilson (eds), 1993, *The Biophilia Hypothesis* (Washington, DC: Island Press).

Kent, Elizabeth, and Leigh Hunt, 1823, *Flora Domestica; or, The Portable Flower Garden* (London: Taylor and Hessey).

Kiley, Dan, and Jane Amidon, 1999, *Dan Kiley in His Own Words: America's Master Landscape Architect* (London: Thames and Hudson).

Kirkham, Pat, 1998, *Charles and Ray Eames: Designers of the Twentieth Century* (Cambridge, MA: MIT Press).

Knowles, Rachel, 2014, 'Sir Richard Colt Hoare, 2nd Baronet (1758–1838)', Regency History (blog), 22 May 2014, www.regencyhistory.net/2014/05/sir-richard-colt-hoare-2nd-baronet-1758.html (accessed 18 March 2020).

Kohlmaier, Georg, and Barna von Sartory, 1990, *Houses of Glass: A Nineteenth-Century Building Type* (Cambridge, MA: MIT Press).

Koppelkamm, Stefan, 1981, *Glasshouses and Winter-Gardens of the Nineteenth Century* (Stuttgart: Verlag Gerd Hatje).

Kowinski, W. S., 1985, *The Malling of America: An Inside Look at the Great Consumer Paradise* (New York: William Morrow).

Krebs, Albin, 1987, 'William C. Pahlmann, Decorator Known for Eclectic Designs, Dies', *New York Times*, 11 November 1987, sec. B, p. 8.

Krohn, Carsten, 2018, *Hans Scharoun* (Basel: Birkhäuser).

Lafuente, Antonio, and Nuria Valverde, 2007, 'Linnaean Botany and Spanish Imperial Biopolitics', in *Colonial Botany: Science, Commerce, and Politics in the Early Modern World*, ed. Londa Schiebinger and Claudia Swan (Philadelphia: University of Pennsylvania Press), pp. 134–47.

Lambert, Phyllis, 2005, '*Stimmung* at Seagram: Philip Johnson Counters Mies van der Rohe', *Grey Room* 20 (Summer), pp. 38–59.

Lamplugh, Anne, 1929, *Flower and Vase: A Monthly Key to Room Decoration* (London: Country Life).

Lancaster, Osbert, 1939, *Homes Sweet Homes* (London: John Murray).

Lange, Christiane, 2006, *Ludwig Mies van der Rohe & Lilly Reich: Furniture and Interiors*, exh. cat. (Ostfildern: Hatje Cantz Verlag).

Lanks, Belinda, and Paul Makovsky, 2009, 'Fifty Years of the Four Seasons', *Metropolis Magazine* (online), 22 July 2009, https://belmontfreeman.com/pdf/2009_07_Metropolis_Lanks_Belina.pdf (accessed 18 March 2020).

Larsen, Christian A. (ed.), 2015, *Philodendron: From Pan-Latin Exotic to American Modern*, exh. cat. (Miami Beach: Wolfsonian-Florida International University).

Latour, Bruno, 2004, *Politics of Nature: How to Bring the Sciences into Democracy* (Cambridge, MA: Harvard University Press).

Leavitt, Sarah A., 2002, *From Catherine Beecher to Martha Stewart: A Cultural History of Domestic Advice* (Chapel Hill: University of North Carolina Press).

Lees-Maffei, Grace, 2001, 'From Service to Self-Service: Advice Literature as Design Discourse', *Journal of Design History* 14, no. 3, pp. 187–206.

Leet, Stephen, 2004, *Richard Neutra's Miller House* (New York: Princeton Architectural Press).

Leitner, Bernhard, 2001, *The Wittgenstein House* (New York: Princeton Architectural Press).

Lewis, Charles A., 1992, 'Foreword' to *The Role of Horticulture in Human Well-Being and Social Development*, ed. D. Reff (Portland, OR: Timber Press), pp. 11–12.

—, 1996, *Green Nature / Human Nature: The Meaning of Plants in Our Lives* (Urbana and Chicago: University of Illinois Press).

Lind, Carla, 2004, *The Wright Style: The Interiors of Frank Lloyd Wright* (London: Thames and Hudson).

Linn, Karl, 2007, *Building Commons and Community* (New York: New Village Press).

Littlejohns, Idalia B., 1927, *Ornamental Homecrafts* (London: Sir Isaac Pitman & Sons).

Logan, Thad, 2001, *The Victorian Parlour* (Cambridge: Cambridge University Press).

Lohr, Virginia I., and Caroline H. Pearson-Mims, 2000, 'Physical Discomfort May Be Reduced in the Presence of Interior Plants', *HortTechnology* 10, no. 1, pp. 53–8.

Lohr, Virginia I., Caroline H. Pearson-Mims and Georgia K. Goodwin, 1996, 'Interior Plant May Improve Worker Productivity and Reduce Stress in a Windowless Environment', *Journal of Environmental Horticulture* 14, no. 2, pp. 97–100.

Long, Christopher, 2002, *Josef Frank: Life and Work* (Chicago: University of Chicago Press).

Loudon, Jane C., 1847, *The Amateur Gardener's Calendar* (London: Longman, Brown, Green and Longmans).

—, 1848, *The Ladies' Flower Garden of Ornamental Greenhouse Plants* (London: William Smith).

—, 2013 [1840], *Instructions in Gardening for Ladies* (London: Constable & Robinson).

—, 2013 [1842], *Botany for Ladies* (London: Books on Demand).

Loudon, John C., 1817, *Remarks on the Construction of Greenhouses* (London: Architectural Library).

—, 2015 [1824], *The Greenhouse Companion* (London: Andesite Press).

Louv, Richard, 2011, *The Nature Principle: Human Restoration and the End of Nature-Deficit Disorder* (Chapel Hill, NC: Algonquin Books).

MacKenzie, John M., 1990, *Imperialism and the Natural World* (Manchester: Manchester University Press).

Maitland, Barry, 1990, *The New Architecture of the Retail Mall* (New York: Van Nostrand Rheinhold).

Makinson, Randell L., and Thomas A. Heinz, 2004, *Greene and Greene: Creating a Style* (Layton, UT: Gibbs Smith).

Maling, E. A., 1862a, *Indoor Plants and How to Grow Them, for the Drawing Room, Balcony and Greenhouse* (London: Smith, Elder and Co.).

—, 1862b, *Flowers for Ornament and Decoration and How to Arrange Them* (London: Smith, Elder and Co.).

Manaker, George H., 1981, *Interior Plantscapes: Installation, Maintenance and Management* (Upper Saddle River, NJ: Prentice Hall).

March, T. C., 1862, *Flower and Fruit Decoration, with Some Remarks on the Treatment of Town Gardens* (London: Harrison).

Marder, Michael, Gianni Vattimo and Santiago Zabala, 2013, *Plant-Thinking: A Philosophy of Vegetal Life* (New York: Columbia University Press).

Mariani, John, and Alex von Binder, 1999, *The Four Seasons: A History of America's Premier Restaurant* (New York: Smithmark).

Martin, Tovah, 1988, *Once Upon a Windowsill: A History of Indoor Plants* (Portland, OR: Timber Press).

Marx, Leo, 1964, *The Machine in the Garden: Technology and the Pastoral Ideal in America* (Oxford: Oxford University Press).

McCoy, Esther, 1977, *Case-Study Houses, 1945–1962* (Santa Monica, CA: Hennessey + Ingalls).

McKechnie, Claire, and Emily Alder, 2012, 'Introduction: Literature, Science, and the Natural World in the Long Nineteenth Century', *Journal of Literature and Science* 5, no. 12, pp. 1–4.

McQuade, Walter, 1964, 'The Ford Foundation's Mid-Manhattan Greenhouse', *Fortune* (October), pp. 177–8.

McQuaid, Matilda, 1996, *Lilly Reich: Designer and Architect* (New York: Museum of Modern Art).

Meeker, Joseph W., 1988, *Minding the Earth: Thinly Disguised Essays on Human Ecology* (Alameda, CA: Latham Foundation).

Merchant, Carolyn, 1980, *The Death of Nature: Women, Ecology and the Scientific Revolution* (San Francisco: Harper).

—, 2005, *Radical Ecology: The Search for a Livable World* (New York and Oxford: Routledge).

Mollinson, John R., 1877, *The New Practical Window Gardener* (London: Groombridge and Sons). [A new edition was published by Henry J. Drane in 1894.]

Moore, Rowan, 2015, 'Walkie Talkie Review – Bloated, Inelegant, Thuggish: 20 Fenchurch Street, London EC3', *The Observer* (*The Guardian* online), 4 January 2015, www./theguardian.com/artanddesign/2015/jan/04/20-fenchurch-street-walkie-talkie-review-rowan-moore-sky-garden (accessed 18 March 2020).

Morton, Timothy, 2007, *Ecology Without Nature: Rethinking Environmental Aesthetics* (Cambridge, MA, and London: Harvard University Press).

Mozingo, Louise, 2011, *Pastoral Capitalism: A History of Suburban Corporate Landscapes* (Cambridge, MA: MIT Press).

Munro, John M., 1971, *The Royal Aquarium: Failure of a Victorian Compromise* (Beirut: American University of Beirut Press).

Muschamp, Herbert, 1996, 'A Modernist Steps out of the Shadows', *New York Times Architectural Review*, 9 February 1996, sec. C, p. 30.

Muthesius, Stefan, 2009, *The Poetic Home: Designing the 19th-Century Domestic Interior* (London: Thames and Hudson).

Myers, William, 2014, *Bio Design: Nature, Science, Creativity* (London: Thames and Hudson).

Nelson, George, 1949, 'Modern Decoration', *Interiors* 109, no. 4 (November), pp. 68–75.

Nerdinger, Winfried, 2019, *Walter Gropius: Architekt der Moderne, 1883–1969* (Munich: Beck C. H.).

Neuhoff, Christine, 2009, 'The Villa Tugendhat by Mies van der Rohe', *International Journal of Architectural Theory* 13, no. 1, www.cloud-cuckoo.net/journal1996-2013/inhalt/en/issue/issues/108/Neuhoff_Tugendhat/neuhoff_tugendhat.php (accessed 18 March 2020).

Neumann, Dietrich, 2007, 'The Tugendhat House: Mies van der Rohe's Czech Masterpiece', *Journal of the Society of Architectural Historians* 6, no. 1, pp. 131–5.

Neutra, Richard, 1935, 'The New Art of Building in California', *California Arts and Architecture* (January).

—, 1954, *Survival Through Design* (Oxford: Oxford University Press).

New York Times, 1985, 'Everett L. Conklin, 77; Created Floral Designs', 22 March 1985, sec. B, p. 5.

Nichols, Kate, and Sarah Victoria Turner, 2017, *After 1851: The Material and Visual Culture of the Crystal Palace of Sydenham* (Manchester: Manchester University Press).

Nierhaus, Irene, 2009, 'The Modern Interior as a Geography of Images, Spaces and Subjects: Mies van der Rohe's and Lilly Reich's Villa Tugendhat, 1928–1931', in *Designing the Modern Interior: From the Victorians to Today*, ed. Penny Sparke, et al. (Oxford and New York: Berg), pp. 107–18.

Nieswander, Judith A., 2008, *The Cosmopolitan Interior: Liberalism and the British Home, 1870–1914* (New Haven and London: Yale University Press).

Olson, Elizabeth, 2006, 'So You Want Ivy Round Your Desk?', *New York Times*, 19 November 2006, sec. 12, p. 2.

Online Archive of California, n.d., entry for Max and Rita Lawrence Architectural Pottery Records, ca. 1950–1994, Online Archive of California (website), www.oac.cdlib.org/findaid/ark:/13030/kt0d5nb767/ (accessed 18 March 2020).

Overy, Paul, 2007, *Light, Air and Openness: Modern Architecture between the Wars* (London: Thames and Hudson).

Pallasmaa, J., 2003, 'Rationality and Domesticity', in *The Aalto House, 1935–36*, ed. Juhani Pallasmaa, Alvar Aalto Architect 6 (Helsinki: Alvar Aalto Foundation), pp. 56–81.

Palmer, Isabelle, 2014, *The House Gardener: Ideas and Inspiration for Indoor Gardens* (London: CICO Books).

Park, Seong-Hyun, and Richard H. Mattson, 2009, 'Therapeutic Influences of Plants in Hospital Rooms on Surgical Recovery', *Horticultural Science* 44, no. 1, pp. 102–5.

Parsons, Russ, Roger S. Ulrich and Louis G. Tassinary, 1994, 'Experimental Approaches to the Study of People–Plant Relationships', *Journal of Home and Consumer Horticulture* 1, no. 4, pp. 347–72.

Pelkonen, Eeva-Liisa, 2011, *Kevin Roche: Architecture as Environment* (New Haven and London: Yale University Press).

Perkins, John, 1877, *Floral Decorations for the Table* (London: Wyman & Sons).

Phillips, B., 1875, 'The Royal Aquarium and Summer and Winter Gardens', *The Times*, 9 October 1875, p. 5.

Piercy, Harold, 1989, *Constance Spry: Creative Ideas in Floristry and Flower Arranging* (London: Christopher Helm).

Pile, John, 1969, 'The Nature of Office Landscape', *AIA Journal* (July), pp. 40–48.

Pimlott, Mark, 2007, *Without and Within: Essays on Territory and the Interior* (Delft: Delft University of Technology).

—, 2016, *The Public Interior as Idea and Project* (Heijningen: Jap Sam Books).

Plumwood, Val, 1993, *Feminism and the Mastery of Nature* (London: Routledge).

Polan, Judy, 2008, 'Villa Tugendhat: A Turbulent History', *Modernism* (Summer), p. 112.

Polin, Giacomo, 1982, *La Casa Elettrica di Figini e Pollini* (Milan: Officina).

Pommer, Richard, and Christian F. Otto, 1991, *Weissenhof 1927 and the Modern Movement in Architecture* (Chicago: University of Chicago Press).

Portman, John, and Jonathan Barnett, 1976, *The Architect as Developer* (New York: McGraw Hill).

Portman, John, et al., 1997, *John Portman: An Island on an Island* (Milan: L'Arca).

Potter, Terry, 1999, *Reflections on Blackpool* (Blackpool: Book Clearance Centre).

Preston, Julieanna (ed.), 2008, 'Interior Atmospheres', special issue, *Architectural Design* 78, no. 3 (May/June), pp. 194–211.

Preston, Rebecca, 1999, '"The Scenery of the Torrid Zone": Imagined Travel and the Culture of Exotics in Nineteenth-Century British Gardens', in *Imperial Cities: Landscape, Display and Identity*, ed. Felix Driver and David Gilbert (Manchester: Manchester University Press), pp. 194–211.

Progressive Architecture, 1968, 'Bürolandschaft USA', May 1968, pp. 174–7.

Proscod, Fred, 2012, *Interior Plantscaping: Reference and Study Manual* (Herndon, VA: Planet).

Qin, Jun, et al., 2013, 'The Effect of Indoor Plants on Human Comfort', *Indoor and Built Environment* 23, no. 5, pp. 709–23.

Raanaas, Ruth Kjaersti, Gret Grindal Patil and Terry Hartig, 2010, 'Effects of an Indoor Foliage Plant Intervention on Patient Well-Being during a Residential Rehabilitation Program', *Horticultural Science* 45, no. 3, pp. 387–92.

Randolph, Cornelia J., 1861, *The Parlor Garden: A Treatise on the House Culture of Ornamental Plants* (Boston: J. E. Tilton).

Rappaport, Erika, 2000, *Shopping for Pleasure: Women in the Making of London's West End* (Princeton, NJ: Princeton University Press).

'Räume voller Grün / Grown Up', 2016, *Form: Design Magazine* 267 (September/October), pp. 14–17.

Rayner, Gordon, 2011, 'Royal Wedding: Kate Middleton to Walk down Avenue of Trees inside Westminster Abbey', *The Telegraph* (online), 26 April 2011, www.telegraph.co.uk/news/uknews/royal-wedding/8474197/Royal-wedding-Kate-Middleton-to-walk-down-avenue-of-trees-inside-Westminster-Abbey.html (accessed 18 March 2020).

Rees, Yvonne, 1995, *Essential English Country Style* (London: Ward Lock).

Reif, Rita, 1972, 'Decorating with Plants', *New York Times*, 5 April 1972, p. 50.

Relf, Diane (ed.), 1992, *The Role of Horticulture in Human Well-Being and Social Development: A National Symposium* (Portland, OR: Timber Press).

Repton, Humphry, 1816, *Fragments on the Theory and Practice of Landscape Gardening* (London: T. Bensley & Sons).

Reyes, Anna, 2017, 'The Benefits of Indoor Plants in Airports', Ambius (website), 18 January 2017, www.ambius.com/blog/indoor-plants-enhancing-the-airport-experience/ (accessed 18 March 2020)

Reynolds, Jim, 1997, '"Palm Trees Shivering in a Surrey Shrubbery": A History of Subtropical Gardening', *Principes* 41, no. 2, pp. 74–83.

Riani, Paolo, and Paul Goldberger, 1990, *John Portman* (Milan: L'Arca).

Rice, Charles, 2016, *Interior Urbanism: Architecture, John Portman and Downtown America* (London: Bloomsbury).

Riezler, Walter, 1931, 'Das Haus Tugendhat in Brünn', *Die Form* 6, no. 9, pp. 321–32.

Robinson, William, 1869, *Plants, Promenades and Gardens of Paris* (London: John Murray).

Rockwell, F. F., and E. C. Grayson, 1947, *The Complete Book of Flower Arrangement* (New York: Windmill Press).

Roncevich, Tim, and Steven Primm, 2011, *Start Your Own Interior Plantscaping Business* (n.p.: CreateSpace Independent Publishing Platform).

Rosner, U., 2008, 'Schminke House in Löbau by Hans Scharoun: To Restore a Major Work of Classic Modern', *The Conservation* 59, no. 1, pp. 11–13.

Rumpfhuber, Andreas, 2011, 'The Legacy of Office Landscaping: SANAA's Rolex Learning Centre', *IDEA Journal*, pp. 20–32.

Ruskin, John, 1907 [1864], 'Of Queen's Gardens', in *Sesame and Lilies: Two Lectures Delivered at Manchester in 1864*, reprint edition (London: George Allen), pp. 87–143.

Sadar, John Stanislav, 2016, *Through the Healing Glass: Shaping the Modern Body through Glass Architecture 1925–35* (London and New York: Routledge).

Saito, Yuriko, 1985, 'The Japanese Appreciation of Nature', *The British Journal of Aesthetics* 25, no. 3, pp. 239–51.

Sanderson, Kenneth, 2002, *Interior Plantscaping* (New York: Delmar).

Saval, Nikil, 2014, *Cubed: A Secret History of the Workplace* (New York: Anchor Books).

Saxon, Richard, 1983, *Atrium Buildings: Development and Design* (London: Architectural Press).

Scheiberg, Susan, 1990, 'Emotions on Display: The Personal Decoration of Work Space', *The American Behavioural Scientist* 33, no. 3, pp. 330–38.

Schiebinger, Londa, 2004, *Plants and Empire: Colonial Bioprospecting in the Atlantic World* (Cambridge, MA: Harvard University Press).

—, and Claudia Swan (eds), 2005, *Colonial Botany: Science, Commerce, and Politics in the Early Modern World* (Philadelphia: University of Pennsylvania Press).

Schivelbusch, Wolfgang, 1992, *Tastes of Paradise: A Social History of Spices, Stimulants and Intoxicants* (New York: Pantheon Books).

Schulze, Franz, and Edward Windhorst, 2012, *Mies van der Rohe: A Critical Biography*, revised edition (Chicago: University of Chicago Press).

Seaton, Beverly, 1995, *The Language of Flowers: A History* (Charlottesville: University Press of Virginia).

Serraino, Pierluigi, 2000, *Julius Shulman: Modernism Rediscovered* (Cologne: Taschen).

Sharples, Joseph, 2007, *Merchant Palaces: Liverpool and Wirral Palaces Photographed by Bedford Lemere & Co.* (Liverpool: Bluecoat Press).

Shaw, Albert, 1890, 'London Polytechnics and People's Palaces', *The Century Illustrated Monthly Magazine* 60 (June), pp. 163–82.

Shephard, Sue, 2010, *The Surprising Life of Constance Spry* (London: Macmillan).

Shibata, Seiji, and Naoto Suzuki, 2002, 'Effects of the Foliage Plant on Task Performance and Mood', *Journal of Environmental Psychology* 22, no. 3, pp. 265–72.

Shulman, Julius, 1977, *The Photography of Architecture and Design* (New York: Whitney Library of Design).

Siebel, Frank, 2017, *Leuchtendes Schiff in schwerer See: Das Haus Schminke im Spiegel des zwanzigsten Jahrhunderts / Kelly Ship in Heavy Seas: The Schminke House in the Mirror of the Twentieth Century* (Zittau: Edition Sächsische Zeitung).

Silva, Valerie, 2017, 'Outside, In: Airports Sprout Indoor Gardens to Help Ease Travel Stress', Apex (website), 1 March 2017, https://apex.aero/2017/03/01/airports-indoor-gardens-ease-travel-stress (accessed 18 March 2020).

Simons, Paul, and John Ruthven, 1995, *Potted Histories: How to Make House Plants Feel at Home* (London: BBC Books).

Sky Garden, n.d., Sky Garden (website), https://skygarden.London/sky-garden (accessed 18 March 2020).

Smith, Elizabeth A. T., 2006, *Case-Study Houses: 1945–1966: The Californian Impetus* (Cologne: Taschen).

Snyder, Stuart D., 1995, *Environmental Interiorscapes: A Designer's Guide to Interior Plantscaping and Automated Irrigation Systems* (New York: Whitney Library of Design).

Soper, Kate, 1995, *What is Nature?: Culture, Politics and the Non-human* (Oxford: Blackwell).

Sparke, Penny, 1995, *As Long as It's Pink: The Sexual Politics of Taste* (London: Pandora).

Sprague, Edward Rand, 1876, *Flowers for the Parlor and Garden* (New York: Hurd and Houghton).

—, 1872, *The Window Gardener* (Boston: Shepard and Gill).

Spry, Constance, 1934, *Flower Decoration* (London: J. M. Dent & Sons).

—, 1951, *Summer & Autumn Flowers* (London: J. M. Dent & Sons).

—, 1952, *How To Do the Flowers* (London: Atlantis).

Stephens, Suzanne, 1978, 'Introversion and the Urban Context', *Progressive Architecture* 59 (December), pp. 27–34.

Stiller, Adolph (ed.), 1999, *Das Haus Tugendhat: Mies van der Rohe, Brünn, 1930* (Salzburg: Anton Pustet).

Strzelecki, Gloria, 2013, '"Make Friends with the Cactus": Floral Art and Australian Modernism', *Australian Garden History* 25, no. 2, pp. 13–16.

Suominen-Kokkonen, Renja, 2007, *Aino and Alvar Aalto – A Shared Journey: Interpretations of an Everyday Modernism* (Helsinki: Lönnbreg Press).

Sutcliffe, G. Lister, 1898, *The Principles and Practice of Modern House Construction* (London: Blackie and Sons).

Syring, Eberhard, and Jörg Kirschenmann, 2007, *Hans Scharoun, 1893–1972: Outsider of Modernism* (Cologne: Taschen).

Talbott, J. A., et al., 1976, 'Flowering Plants as a Therapeutic/Environmental Agent in a Psychiatric Hospital', *Horticultural Science* 50, no. 4, pp. 365–6.

Tifferet, Sigal, and Iris Vilnai-Yavetz, 2017, 'Phytophilia and Service Atmospherics: The Effect of Indoor Plants on Consumers', *Environment and Behavior* 49, no. 7, pp. 814–44.

Tigerman, Bobbye, 2007, '"I Am Not a Decorator": Florence Knoll, the Knoll Planning Unit and the Making of the Modern Office', *Journal of Design History* 20, no. 1, pp. 61–74.

—, 2013, *Handbook of California Design, 1930–1965: Craftspeople, Designers, Manufacturers* (Los Angeles: LACMA).

TLCstaff, 2011, 'Rebirth of a Landmark', Total Landscape Care (website), 1 November 2011, www.totallandscapecare.com/landscaping/rebirth-of-a-landmark/ (accessed 18 March 2020).

Todd, Dorothy, and Raymond Mortimer, 1929, *The New Interior Decoration: An Introduction to Its Principles, and International Survey of Its Methods* (London: Batsford).

Turner, Brad, 2009, 'House in Nagoya by Suppose Design Office', *Dezeen*, 15 July 2009, www.dezeen.com/2009/07/15/house-in-nagoya-by-suppose-design-office/ (accessed 18 March 2020).

Ulrich, Roger S., 1979, 'Visual Landscapes and Psychological Well-Being', *Landscape Research* 4, no. 1, pp. 17–23.

—, 1984, 'View through a Window May Influence Recovery from Surgery', *Science* 224, no. 4647, pp. 420–21.

—, and Russ Parsons, 1992, 'Influences of Passive Experiences with Plants on Individual Well-Being and Health', in *The Role of Horticulture in Human Well-Being and Social Development*, ed. Diane Relf (Portland, OR: Timber Press), pp. 93–105.

Valen, Dustin, 2016, 'On the Horticultural Origins of Victorian Glasshouse Culture', *Journal of the Society of Architectural Historians* 75, no. 4, pp. 403–23.

Veder, Robin, 2007, 'Mother-Love for Plant-Children: Sentimental Pastoralism and Nineteenth-Century Parlour Gardening', *Australasian Journal of American Studies* 26, no. 2, pp. 20–34.

Villa Tugenghat, n.d., 'The Housing Philosophy', Villa Tugendhat (website), www.tugendhat.eu/en/the-building/the-housing-philosophy.html (accessed 18 March 2020).

Walton, John K., 2000, *The British Seaside: Holidays and Resorts in the Twentieth Century* (Manchester: Manchester University Press).

Ward, Nathaniel Bagshaw, 1842, *On the Growth of Plants in Closely Glazed Cases* (London: John Van Voorst).

Warner, Charles F., 1917, *Home Decoration* (London: T. Werner Laurie).

Watson, Bruce, 2016, 'The Troubling Evolution of Corporate Greenwashing', *The Guardian* (online), 20 August 2016, www.theguardian.com/sustainable-business/2016/aug/20/greenwashing-environmentalism-lies-companies (accessed 18 March 2020).

Watson, Rosamund Marriott, 1897, *The Art of the House* (London: George Bell & Sons).

Whittingham, Sarah, 2009, *Fern Mania: The Story of Pteridomania, a Victorian Obsession* (London: Frances Lincoln).

Wilk, Christopher, 1981, *Marcel Breuer: Furniture and Interiors* (New York: Museum of Modern Art).

—, 2006, *Modernism: Designing a New World, 1914–1939* (London: V&A Publications).

Wilkinson, Anne, 2012, *Shirley Hibberd, the Father of Amateur Gardening: His Life and Works, 1825–1890* (Birmingham: Cortex Design).

Williams, Henry T., 1872, *Window Gardening: Devoted Specially to the Culture of Flowers and Ornamental Plants for Indoor Use and Parlor Decoration*, 2nd edition (New York: Henry T. Williams).

Williams, Kevin, 2007, *Seed to Elegance: Kentia Palms of Norfolk Island, South Pacific* (Norfolk Island: Studio Monarch Books).

Williams, Raymond, 1983, *Culture and Society, 1780–1950* (New York: Columbia University Press).

Wilson, Edward O., 1984, *Biophilia: The Human Bond with Other Species* (Cambridge, MA, and London: Harvard University Press).

Wilson, Robert A., 2004, *Epitome of Desire: The Story of the Nashers of Texas and One of the World's Greatest Sculpture Collections, Created by Their Passion for the Best* (Austin: University of Texas Press).

Winter, John, 2000, 'Ludwig Mies van der Rohe: The Tugendhat House', *The Architectural Review* 1243 (September), pp. 16–19.

Wolverton, B. C., 1996, *Eco-Friendly House Plants* (London: Weidenfeld and Nicolson.

—, Anne Johnson and Keith Bounds, 1989, *Interior Landscape Plants for Indoor Air Pollution Abatement* (report), NASA-TM-101766, NAS 1.15:101766 (Bay Saint Louis, MS: John C. Stennis Space Center).

Woo, Elaine, 2010, 'Max Lawrence Dies at 98; Co-founder of L.A.'s Architectural Pottery', *Los Angeles Times* (online), 1 August 2010, www.latimes.com/local/obituaries/la-xpm-2010-aug-01-la-me-max-lawrence-20100801-story.html (accessed 18 March 2020).

Wood, R. A., et al., 2002, 'Potted-Plant / Growth Media Interactions and Capacities for Removal of Volatiles from Indoor Air', *Journal of Horticultural Science and Biotechnology* 77, no. 1, pp. 120–29.

Woods, May and Arete Warren, 1990, *Glasshouses: History of Glasshouses, Orangeries and Conservatories*, reprint edition (London: Aurum Press).

Wright, Patrick, 1985, *On Living in an Old Country: The National Past in Contemporary Britain* (London: Verso).

Wulf, Andrea, 2009, *The Brother Gardeners: Botany, Empire and the Birth of an Obsession* (London: Windmill Books).

—, 2015, *The Invention of Nature: The Adventures of Alexander von Humboldt, the Lost Hero of Science* (London: John Murray).

—, and Emma Gieben-Gamal, 2005, *This Other Eden: Seven Great Gardens and 330 Years of English History* (London: Little, Brown).

Wynn, Christopher, 2017, 'Dallas Architect E. G. Hamilton, Lead Designer of NorthPark Center, Dies at 97', *Dallas Morning News* (online), 9 May 2017, www.dallasnews.com/arts-entertainment/architecture/2017/05/09/dallas-architect-e-g-hamilton-lead-designer-of-northpark-center-dies-at-97/ (accessed 18 March 2020).

Yaneva, Albena, and Alejandro Zaera-Polo, 2015, *What is Cosmopolitical Design? Design, Nature and the Built Environment* (London and New York: Routledge).

Yeang, Ken, 1995, *Designing with Nature: The Ecological Basis for Architectural Design* (New York: McGraw Hill).

Ziegler, Catherine, 2007, *Favored Flowers: Culture and Economy in a Global System* (Durham, NC, and London: Duke University Press).

Zunz, Oliver, 1990, *Making America Corporate, 1870–1920* (Chicago: University of Chicago Press).

Index

Page numbers in *italics* indicate illustrations

A

Aaksen Responsible Architecture: house in Bandung, Indonesia 187, 189
Aalto, Alvar and Aino 12, *80*, 81
Abbott, Ruth *167*
'Action Office' (Herman Miller and Nelson) 173
Adam, Robert 24
agaves *26*, 76, 81, 101
air plants 117
air pollution 136, 173, 178–9
Albert, Prince 54
Alexandra Palace, Muswell Hill 56, 194
Alfred, Prince, Duke of Edinburgh 57, 59
Amazon headquarters, Seattle 9, *10*
American Nurseryman Magazine 168
& Other Stories, Sweden 196
Anderson, Harvey 125
Anne, Queen 16, 24
Anthropologie store, Regent Street, London (Biotecture) 195, 196
Antoine Graves Homes, Atlanta 146–7, 149
aquaria 33, *34*, 35, 46, 56, 57
The Architects Collaborative 164–5
Architectural Pottery 119–20
architecture and architects
 on the benefits of nature inside 170
 interior decorators and designers relationship 126, 128, 129, 139, 141, 153, 158, 165
 landscape architects relationship 118–19, 122, 126, 133, 139, 141
 modernist architecture *see under* modernism
 organic architecture 71, 144
 plantscapers relationship 126, 137, 165–6, 168
 see also landscape architecture
Art Nouveau 66
Arts and Crafts movements
 American 107
 British 66
Arts & Architecture journal 111, 120, 122, 123
Ashburnham Pavilion, Cremorne Gardens 52
aspidistras 10, 35, 69, 185
Associated Landscape Contractors of America 178
Atocha railway station, Madrid 192
Aublet, Jean-Baptiste-Christophe Fusée 22
Australian umbrella trees 117, 151, 153, 159

B

Badovici, Jean *72*
Bailey House (Case Study House No. 21), Hollywood Hills Canyon (Koenig) 120–21, *121*
Balmori, Diana *191*, 192
bamboo 123, *140*, 141, 173, 175
Banks, Sir Joseph 16, 21
Barn, Los Angeles 118, 123
Barnett, Jonathan: *The Architect as Developer* (with Portman) 144, *144*
Barr, Alfred H. (MoMA) 131
Baum, Joe (Restaurant Associates) 128, 129, 130
Beacon Hill House conservatory, Newport, Rhode Island 27
Bednar, Michael J. 143, 148
begonias 197
Bellevue House greenhouses, Ireland 26
benefits of nature inside
 biophilia relationship 168–70, 183
 commercial benefits 11, 13, 15–16, 51, 65, 163–4, 176, 177–8, 179, 183, 190
 emotional benefits 37–8, 45, 49, 145, 171, 183–4, 190, 192
 environmental benefits 12–13, 38–9, 161, 168, 170, 178–9, 181, 186, 190
 health and well-being 66, 107, 117, 161, 163, 168, 170–71, 173, 175, 179, 190, 197
 measuring 170–71, 173, 175, 177–8, 179, 186
 mental health 117, 161, 163, 166, 170–71, 173, 175, 179, 194, 197
 noise-absorbing benefits 173, 179
Berlin winter gardens 52–3
Bijvoet, Bernard 83
Biltmore Hotel, New York 64
Biodome, Montreal 182
biophilia 168–70, 183
Biotecture 168, *195*, 196, 197
Blackpool Winter Gardens 60–61, *61*
Blanc, Patrick 196
Bobbink & Atkins nursery 131
Bochynek, Johann 76
Boeri Studio
 Bosco Verticale ('Vertical Forest'), Milan 189, 190

Trudo Vertical Forest social housing project, Eindhoven 189
Bonar, Ann 184–5
books on gardening and related subjects
  nineteenth century 34, 42, 44, 45, 46–8, 186
  twentieth century 82, 83, 85, 164, 164, 165–7, 166, 186
Botanic Gardens, Belfast 22
Botanical Garden, New York 22
botanical gardens 15, 16, 17, 19, 20–22, 21, 41, 51, 192
  see also Royal Botanic Gardens, Kew
Bourgeois, Elise 83
Bournemouth Winter Gardens 61
Bretton Hall conservatory, West Yorkshire 25
Breuer, Marcel 153
  apartment for Erwin Piscator in Berlin 81, 82
  Doldertal apartment building, Zurich (with Alfred and Emil Roth) 76, 77
  house for Marianne Harnischmacher in Wiesbaden 84
British East India Company 16
Brody House, Los Angeles (Jones) 118
Brody, Jerry (Restaurant Associates) 127, 128, 129
Brown, Lancelot 'Capability' 19, 25
Brown Palace Hotel, Denver 146, 147
Budo Pradono: houses in Jakarta and Depok 189
Burlington, 3rd earl of 19
Burton, Decimus 22, 54

C
cacti 12, 30, 35, 69, 76, 78, 81, 82, 83, 109, 117, 123, 196
Calcutta Botanic Garden, India 21
California see Californian modernism; southern California
Californian modernism 11, 105, 106, 122–3, 125
  affordable housing projects 111
  desert location reflected in planting schemes 109, 110, 117
  pastoral ideal and machine age relationship 107, 109, 111
  private indoor spaces
    cut flowers 109, 110, 117
    furniture and furnishings 109, 110, 120–21, 122, 123
    inside/outside lifestyle 105, 106, 107, 109, 110, 111, 114, 119, 119, 121, 122, 123, 125
  internal flower beds 118, 118–19, 119
  materials used 107, 111, 123
  open plan living 107, 109, 119
  potted plants and planters 106, 113, 114, 115, 116, 117; blurring the boundaries between inside and outside 111–12, 112, 114, 119–20; in internal gardens 118, 118; used as screens 119, 119; used to define spatial relationships 114, 120–21, 121
  role of plants and flowers 107, 122
  trees incorporated 118
  relationship with nature 105, 107, 109, 111, 117
  steel, use of 107, 111, 118, 120, 121
  see also Case Study Houses
Carlton Hotel, London 63
Carson, Rachel: Silent Spring 181
Case Study Houses 111, 118, 118, 119, 120–21, 121, 122
  see also under Eames, Charles and Ray
Castellano, Aldo 145–6, 151
Centennial Exhibition (1876), Philadelphia 56
Central Hotel, Berlin 63
Central Park, New York 23–4
Chang Architects 189
Changi airport, Singapore (Peter Walker and Partners) 196, 197
Chareau, Pierre 83
Chatsworth House conservatory, Derbyshire 25, 54
Chavalliaud, Léon-Joseph 23
Chelsea Physic Garden 21
Chippendale, Thomas and Chippendale style 33, 40
chrysanthemum house, Battersea Park, London 23
chrysanthemums 39, 128, 135, 151, 153, 158, 175
Clark, Frank 22–3
Cole, Henry 54
commercial nurseries 15, 16, 19, 20, 22, 126, 164, 167–8
commercialism
  benefits of nature inside 11, 13, 15–16, 51, 65, 163–4, 176, 177–8, 179, 183, 190
  environmentalism exploited 135, 182
  plantscaping relationship 173, 176, 177
  public indoor spaces relationship 49, 51, 56, 57, 59–60, 65, 126, 131, 144
Conklin, Everett Lawson 130–31, 137, 139, 141, 164–5, 168–9, 173, 175, 179
Connolly, Shane 192
conservatories
  attached to houses 19, 25–7, 28, 29, 33
  choice of plants and flowers 11, 29–30, 31
  for the middle classes 19, 29, 29, 30, 30, 33, 99
  in modernist houses 76, 77
  as public spaces 21, 22, 23–4, 24, 25, 30–31
  see also glasshouses; winter gardens
Conservatory of Flowers, Golden Gate Park, San Francisco 23, 24
containers 9, 35, 81, 83, 119–20, 137, 158, 172
  see also planters; Wardian cases; Warrington cases
Cremorne Gardens, Chelsea 52, 52, 57
Crown Sky Garden, Ann & Robert H. Lurie Children's Hospital, Chicago (Kim) 175
Crystal Palace, Hyde Park 24, 29, 54–5, 75, 194
Crystal Palace, Sydenham 55, 55
Cunard 64
cycads 97

D
Dalbet, Louis 83
Dalkeith Park conservatory, near Edinburgh 25
Dalsace, Jean 83
Darwin, Charles 37
David Rubenstein Atrium, Lincoln Center, New York 197
Davidson, O. Wesley 130
De Sandalo, Rudolf 90, 93, 95
Destruction of the Country House, 1875-1975 exhibition (1974), Victoria & Albert Museum 184
Deutscher Werkbund exhibition (1927), Stuttgart 71, 74, 100
Dickinsons' Comprehensive Pictures of the Great Exhibition of 1851 53
Dinkeloo, John 136, 136
Djo-Bourgeois, Georges 83
Docker, Richard 71
dracaenas 35, 39, 83, 159, 172, 173
Dunlap Psychiatric Hospital, New York 175
Dupont offices, Wilmington, Delaware 173
Dutch East and West India Companies 16–17

Index 217

E

Eames, Charles 123
  Case Study House No. 9, Pacific Palisades 119
Eames, Charles and Ray 12
  apartment, Strathmore Apartments, Los Angeles (Neutra) 114
  Case Study House No. 8, Venice, CA 111–12, *112, 113*, 114, *114, 115*, 117
  furniture 131
Earl Swensson Associates *177*
Eckbo, Garrett 118, 122, 123, 137
eclecticism in the nineteenth century *see* historicism and eclecticism
Edbrooke, Frank E. 147
Eden Project, Cornwall *182*, 182–3
Edris House, Palm Springs 119, *119*
education and nature inside 22, 33, 37, 49, 54, 56, 57, 59, 65, 103, 183
Edward Larrabee Barnes & Associates *140*
Edwards, H. Griffith (Edwards & Portman Architects) 143
Electric House for Fourth International Triennial Exhibition (1930) Monza (Group 7) 78, *79*
Elok House, Singapore (Chang Architects) 189
Embarcadero hotel, San Francesco (Portman) 156
Enlightenment 17, 37, 67
Entelechy I and II (Portman) 144–6, 151
environmentalism and nature inside
  benefits 12–13, 38–9, 161, 168, 170, 178–9, 181, 186, 190
  commercialism exploits 135, 182
  environmental crises 12, 13, 181, 184, 197
  environmental movements 11, 12, 13, 167, 179, 181–2
  scepticism 182, 183
Entenza, John 111
Etsy headquarters, Dumbo, Brooklyn 197
Evans, Michael R. 176
Evelyn, John
  greenhouse 24
  Kalendarium Hortens 24
Everett Conklin & Company International 131
exhibition buildings 51, *53*, 55–6
exotic plants 16, 19–20, 29–30, 37, 41, 53–4, 55, 123, 125, 126, 164
  see also trade in exotic plants

F

fashion in nature inside 11, 12, 26, 29, 30, 34, 49, 84–5, 166, 179
fatsias 120, 173
ferns 37, 41–2
  growing cases *34*, 35, 47
  popularity 42, 45, 185
  in private indoor spaces *30*, 31, 35, *36*, 39, *41*, 42, 97, *106*, 121
  in public and indoor spaces 139, *183*, 197
Figini, Luigi (Group 7) 78
figs and fig trees 83, 128, *137*, 173, 178, 185, *193*
flower arranging
  books 47–8, 85
  gender issues 45, 47, 85
  as interior decoration 45, 47–8, 49, 85
  Japanese (*ikebana*) style 48, 83–5
  in modernist houses 83–5, 90, 95, 99–100, *100*, 101
  nineteenth century 45, 47–8, 49
Follis, John (Architectural Pottery) 120
Ford Foundation Building, New York 135–7, 139, 149
  indoor garden 131, 135–6, *136*, 137, *137, 138*, 139, *139*; lighting 139; planters 137; pool 136, 137, *138*; staircase 137, *137*; trees and shrubs *136*, 137, *137*, 139, *139*
  materials used 135, 136
Die Form journal 102
Fortune, Robert 20
Forum of the Twelve Caesars, Rockefeller Center, New York 128
Foster + Partners
  Apple store, Regent Street, London *193*
  Apple store, Singapore 192
  Crossrail Place, Canary Wharf, London *193*, 194
  Oslo railway station 194
Four Seasons Restaurant, New York 126–31, *128*
  four seasons theme 130
  furniture and furnishings 131
  Grill Room 127, 131
  lighting and temperature 130
  Pool Room 127, *128*
  potted plants 127, *128*
  sculpture and other artworks 131
  tableware 131
  trees in pots 127–8, *128*, 130, 191
Fourth International Triennial Exhibition of Modern Decorative and Industrial Arts, Monza (1930) 78

Frank, Josef 76, 84
Frederick the Great, King of Prussia 24, 25
Frey, Albert 106, 117, 123
Friedman, Alice T. 109–10
Fromm, Erich 168
functionalism and functionality 67, 69, 70, 141
Furuta, Tok: *Interior Landscaping 166*, 167, 168

G

G-Tower, Incheon, South Korea (HAEAHN Architecture) *188*, 189
Gaines, Richard L.: *Interior Plantscaping* 165–7, 168
Gamble House, Pasadena, California (Greene & Greene) 74, 107
Garcia House, Los Angeles (Lautner) 117
Gardner, Isabella Stewart 76
  conservatory in Boston house 27
Gate 25, Terminal 3, Heathrow Airport 197
gender issues
  flower arranging 45, 47, 85
  gardening indoors and outdoors 11–12, 26, 27, 29, 42, *42*, *43*, 45–6, 47–9
  in interior decoration and design 12, 34, 46
  women in the workplace 164, *165*
geraniums 17, 37, 111
Gesner, Harry 117
Giedion, Sigfried 76, *77*
Gilliatt, Mary 186
Gillespies 194, *194*
Girard, Alexander 137
glass-and-iron buildings 25, 54, 55, 56, 66
  see also Crystal Palace
glasshouses 22, 24, 29, 30, 31, 51–2, 55, 56, 183
  see also conservatories; greenhouses
Gordon, Elizabeth (*House Beautiful*) 130
Graf, Alfred B.: *Exotic House Plants* 164, *164*
Graf Zeppelin airship (Pankok) 84, *84*
Grand Central Railway Terminal, New York 64
The Grange conservatory, Hampshire 27
grasses 123, 135, 139
Gray, Eileen *72*
Great Exhibition (1851) 24, *53*, 54
Great London Exposition (1862), South Kensington 56
Great Yarmouth Winter Gardens 61, *62*, 63

218 Index

green walls *see* living walls
Greenbriar Shopping Center, Atlanta 146
Greene, Charles and Henry (Greene & Greene) 74, 107, 151
greenhouses 12, 21, 24–5, 26, 31
see also glasshouses
Gropius, Walter 84
Auerbach House, Jena, Germany 76
Masters' House, Dessau, Germany *74, 75*
Zuckerkandl House, Jena, Germany 76, *77*
Group 7 (Figini and Pollini) 78, *79*
Gruen, Victor 132
Guerra, David 189

H
Hackney Botanic Nursery (Loddiges family) 40, 42, 55
HAEAHN Architecture *188*, 189
Haines, William 118
Halprin, Lawrence *132*, 133, 135
Hamilton, E. G. 133
Hammer, Nelson 164–5
Hammerbacher, Herta 71
Hampton Court, London: glasshouse and orangery 24
hanging baskets 35, *44*, 47, 48
Häring, Hugo 71
Harrods, London 65
Hartl, H.: *Modern Interiors in Colour*: 'Living Room with Cacti' *82*
Hassard, Annie: *Floral Decorations for the Dwelling House* 48
Haus & Garten (Frank and Wlach) 76
Hawaiian Room, Hotel Lexington, New York 128
Heath, Oliver (Oliver Heath Design) 168, *169*
Hepworth, Barbara: *Helicoids in Sphere* sculpture, Doldertal apartments, Zurich 76, *77*
Herman Miller 173
Hibberd, Shirley
Rustic Adornments for Homes of Taste 26, 33, 41, 46, 47, 184; 'A Modification of the India Table Decoration' *40*, 41; 'Conservatory Leading from Dwelling-House' *28*
historicism and eclecticism in the nineteenth century 10, 40–41, 61, 66
history and continuity of nature inside 9–11, 12, 13, 15, 87, 103
Hoare, Henry, II *18*
Hoare, Richard Colt 33

Hoffman, Herbert: *Modern Interiors in Europe and America* 83
The Home of Today (*Daily Express* book) 85
Hooker, Sir Joseph 41
horticulturists 15, 16, 17, 20, 22, 126, 130–31, 136
Hortus Botanicus, Amsterdam: Palm House 22
Hortus Botanicus, Leyden 22
Hotel Fontainebleau, Miami (Lapidus) 148
Hotel Hermitage, Monte Carlo 63
hotels in twentieth-century late modernism 125, 143, 148, 156
see also Hyatt Regency Hotel
hothouses 24, 30
see also glasshouses
House II, Palm Springs (Frey) 117
Hunt, Leigh: *Flora Domestica; or, the Portable Flower Garden* (with Elizabeth Kent) 46
Huxtable, Garth and Ada Louise 131, 139
Hyatt Regency Hotel, Atlanta 141, 148–51, *152*, 153–4, *155*, 156, *157*, 158–9, 161
atrium 148, 149, *150*, 150–151, *152*, 153, 154, *155*, *157*; aviary 148, 149, 156, 161; balconies *150*, 151, *157*, 158–9; colour 148, 156, 158, 159, 161; furniture 153; materials used 153, 154, 156; plants and planters 151, 153, 156, *157*, 158–9, 161; seating area *152*, 153; water features 153, 154, 156, 161
criticised as 'inward looking' 149
entrance tunnel 150
glass-walled elevators *152*, 153, 154, *155*
'Kobenhavn Kafe' (Copenhagen Café) 153, *154*
lighting 156
Parasol Lounge with plexiglass 'parasol' 154, *154*, 156, 158
Polaris revolving rooftop restaurant 154, 156
postcard *159*
Terrace Floor: Hugo's Gourmet Restaurant 156

I
IBM Building, New York *140*, 141
industrialisation 11, 22, 34, 135
Interface Showroom for Clerkenwell Design Week (2016), 168
interior decorators and designers

architects relationship 126, 128, 129, 139, 141, 153, 158, 165
on the benefits of nature inside 170
flower arranging 45, 47–8, 49, 85
gender issues 12, 34, 46
modernism and late modernism 70, 81, 125, 126, 129–30, 153
nostalgia influences 154, 184–5, *185*, 186, 189
plantscapers relationship 126, 137, 165–6, 168
role in nature inside 10, 34, 118–19, 125, 126, 130, 158, 181, 191
Interior Design magazine 125, 130
interior landscaping *see* plantscaping
Isenstadt, Sandy 73
Isla Gladstone Conservatory, Stanley Park, Liverpool 23
ivy 34, 35, *35*, 39, 74, 81, 146, 173, 184, 197

J
Jackson and Lincoln parks conservatories, Chicago 24
James, William Curtiss 27
Japan
architecture and architects 186–7
flower arranging style (*ikebana*) 48, 83–5
nature and culture debate 186–7
The Japanese House exhibition (2017), Barbican, London *187*
Jardin des Plantes, Paris: Mexican Hothouse 22
Jardin d'Hiver, Paris 53–4
Jardines del Retiro, Madrid: crystal palace 55
John Hancock Child Care Center, Boston 175
Johnson, Louisa: *Every Lady Her Own Flower Gardener* 42, 45
Johnson, Philip 127, *127*, 128, 129, 130
Jones, A. Quincy 111, 117, 118–19, 123
Julio Hernández House, Morelia, Mexico (Roof Arquitectos) 189
Jutras, Roland (Roland Wm. Jutras Associates Inc.) 153, 154

K
Kahn, Louis 144
Kaplan, Rachel and Stephen 170–71, 173, 175
Kaufmann, Edgar J. 73, 109
Kellert, Stephen: *The Biophilia Hypothesis* (with Wilson) 170

Kelly, Richard 130
Kent, Elizabeth: *Flora Domestica; or, the Portable Flower Garden* (with Hunt) 46
Kent, William 19
Kentia palms 39, 41, 164, 173
Kew Gardens *see* Royal Botanic Gardens, Kew
Khoo Teck Puat hospital, Singapore 175
Kiley, Dan 136–7, 139, 165
Kim, Mikyoung 175
King George VI Memorial Park conservatory, Ramsgate, Kent 25, *26*
Knockbreda health centre, Belfast 192
Koenig, Pierre 120–21, *121*

L

La Orotava Botanic Garden, Tenerife 21
Lambert, Phyllis (Seagram) 127, *127*, 128, 129
Lamplugh, Anne 85
landscape architecture and architects 118, 130, 133, 136–7, 164–5, 194
  architects relationship 118–19, 122, 126, 133, 139, 141
  on the benefits of nature inside 170
  plantscapers relationship 130, 141, 164–5
landscape gardening and design 17, 19, 22, 25, 29, 105–6
  see also landscape architecture
Lang, Anna and Robert 76
Langham Hotel, London 63
Lapidus, Morris 148
late modernism 123, 141
  private indoor spaces 106
  public indoor spaces; blurring of interior and exterior boundaries 145, 146; exterior features 145, 146; illusion of domesticity 125, 126, 130, 132–3, 143, 145, 147, 148, 149, 161; importance of nature; inside 12, 15, 16, 130, 141, 143, 145, 146; interior decoration and design 125, 126, 129–30, 153; planting 145, 146; steel, use of 135; water features 145; *see also under* public and semi-public indoor spaces
Latour, Bruno 13
Lautner, John *116*, 117
Lawrence, Max and Rita (Architectural Pottery) 119–20
Le Corbusier 106, 191
  Cabanon, Roquebrune-Cap-Martin, France 73
  Curutchet House, Argentina 75–6

Esprit Nouveau pavilion, International Exhibition (1925), Paris 75
Mill Owners' Association Building, Ahmedabad, India 76
Villa Savoye, Poissy, France 71, 73, 87
Villa Stein, Paris 74
Leet, Stephen (*Architectural Record*) 109
Lehrer's Flowers *167*
leisure spaces in the nineteenth century 51, 52, *52*, 54–7, *58*, 59–60, 66, 194
  other facilities 55, 57, 59
  for working-class populations 22, 52, 57, 59–60
Linn, Karl 130
Linnaeus, Carl 17, 23, 163
living walls 168, *169*, *195*, 196–7
Loddiges family 40, 42, 55
Loudon, Jane: *Instructions for Gardening for Ladies* 42, 45
Loudon, John 20, 25, 27, 29, 53
  Encyclopaedia of Gardening 52–3
Louis Stern-Hugo Reisinger mansion conservatory, 993 Fifth Avenue, New York 27
Lovell, Philip 106, 107

M

Maison de Verre, Paris (Chareau) 83
Maling, Miss A. E.
  fern case 47
  Flowers for Ornament and Decoration 47–8
Mallet-Stevens, Robert 75
Malone, Hollis (Opryland Hotel) 177
Manaker, George H. 125, 167, 168
Manning, Warren 137
maples 9, 92, 93, *93*
March, Thomas: *Flower and Fruit Decoration* 47
Margate winter garden 63
Marx, Leo 107, 126, 169
Mattson, Richard H. 175
Merchandise Mart, Atlanta 144, 148
Mewès and Davis 63
Mies van der Rohe, Ludwig 91, 110, *127*, 131, 132
  Barcelona chairs 95, 99
  Brno chairs 95, 97, 100, 129
  Exhibition House, 'Die Wohnung unserer Zeit,'exhibition (1930–31), Berlin (with Reich) *100*, 101
  Farnsworth House, Illinois 87
  Glass Room, Deutscher Werkbund exhibition (1927), Stuttgart (with Reich) 100–101

Lake Shore Drive apartments, Chicago 101, *101*
Seagram Building, New York 126–7, 129
Tugendhat chairs and tables 95, *95*
see also Villa Tugendhat
Miller, Grace 107, 109, *110*
Miller House, Indianapolis 137
modernism 11, 67, 105
  modernist houses 69–70, 74; balconies, verandas, terraces, roof gardens and external staircases 73, 74, *74*, 75, 81, 89; conservatories 76, *77*; flower arranging 83–5, 90, 95, 99–100, *100*, 101; habitability issues 87, 102; for middle-class clients 87–8, 90, 92, 102; steel, use of 70, 88, 90, 91, 107; technological innovation 10, 78, 87, 88, 90; for working classes 87
  natural setting important *70*, 71, 73, 75, 144
  private indoor spaces 11, 69–70; blurring of interior and exterior boundaries 67, 71, 74, 75, 76, 81, 85, 94, 97; 'clutter' eliminated 67, 69, 70, 81, 83, 87, 102, 117; double-glazing used as miniature greenhouses 78, *79*, 81; indoor gardens 73, 76, *77*, *80*, 81; interior decorators and designers 70, 81; potted plants 75, 76, 81, *82*, 83, *100*, 101, *101*; role of plants and flowers 11, 12, 70, 71, 76, 87, 102, 110; views framed like artworks on walls 73, 75, 94, 95, 97; winter gardens and conservatories 73, 76, *77*; *see also* Villa Tugendhat
  public indoor spaces; ocean liners 76, 78, *78*; *see also under* public and semi-public indoor spaces
  respect for nature 75–6, 85, 107, 117
  revival in twenty-first century 186–7, *187*, 189, 190
  see also Californian modernism; late modernism
Mollinson, John R.: *The New Practical Window Gardener* 34, *44*, 48, 183
Moore, Charles 41
Moriyama House, Tokyo (Nishizawa) 187, *187*
Mortimer, Raymond 70
Morton, Timothy 181, 184
mother-in-law's tongues 76, 92, 119, *119*
Musée du quai Branly, Paris (Nouvel) 196
Mutual Housing Association (MHA) 111
  site office (Jones) 117
Myrick, Richard B. 133

N

NASA 178
Nash, John 27
Nasher, Ray and Patsy 133
nature and culture debate 11, 13, 40, 73, 89, 107, 109, 146, 169–70, 186–7, 189
Necchi, Gigina and Nedda 78
Nelson, George 173
Netherlands: 'tulipomania' 33
Neutra, Richard 105–6, 107, 109, 111, 114
  Demonstration Health House, Los Angeles 107
  Desert House, Palm Springs 109
  Grace Miller House, Palm Springs 107, *108*, 109, *110*
  The Machine in the Garden: Technology and the Pastoral Ideal in America 107
  on photographing interiors 109–10
  Strathmore Apartments, Los Angeles 114
  Vienna Werkbundsiedlung house 75
New York Crystal Palace 55
Nishizawa, Ryue (Studio SANAA) 187, *187*
NorthPark Center, Dallas 131–3, *132, 134,* 135, 136, 149
  climate control 133
  materials used 135
  Nieman Marcus store 131, *132,* 136
  potted plants *134,* 136
  sculpture and other artworks *134,* 135
  water features *132,* 133, 135
Nouvel, Jean 196

O

Oakland Museum, California 136–7
office buildings in twentieth-century late modernism 125, *140,* 141, 143
  see also Ford Foundation Building
Opryland Hotel, Nashville 176–7, *177*
orangeries 24, *25,* 26, 51, 73
orchids 30
Orton Hall conservatory, Peterborough 26
Osterley House, London: orangery 24
Overy, Paul 71, 75
Oyj, Martela *174*

P

Pahlmann, William 128, *129,* 129–30
Palace Hotel, San Francisco 64, *65*
palm courts 63
  department stores 51, 55, 63, 65
  disappearance of in twentieth century 69
  hotels 51, *62,* 63–4, *64, 65*
  ocean liners 51, 63, 64–5

palm houses 22, 23, *23,* 24, 51
palms
  in private indoor spaces 27, *30,* 31, 35, *38, 39,* 39–40, *40,* 41, 118, *118,* 185
  in public indoor spaces 13, 37, 63, *65,* 65–6, 78, 128, 146, 147, 159, 173, 178, *183, 191,* 191–2; *see also* conservatories; palm courts; palm houses
'Palmyra', Aigburth Vale, Liverpool
  conservatory *30,* 31, 40; ferns and palms *30,* 31, *39,* 40
  dining room *39,* 40–41
Pamplemousses Botanic Garden, Mauritius 21
Pankok, Bernhard 84, *84*
Paris Exhibition (1900) 56
Park, Seong-Hyun 175
Parrott, Aubrey 158
Paxton, Joseph 24, 25, 29, 54
Peabody Hotel, Memphis 148
peace lilies 178, 197
Pelli, César *191,* 191–2
Penda 189
People's Palace, Mile End Road 56, 57, *58*
  facilities 57
  shows and entertainments 59
  winter garden 57, *58,* 59
Perkins, John: *Floral Decorations for the Table* 47
Peter Walker and Partners *196,* 197
Phillips, Bruce 57
philodendrons 128, 151, *157,* 158, *171, 172,* 173
physic gardens 20–21
plant-hunters and collectors 15–16, 20, 21
plant transportation boxes *see* Wardian cases
Planterra 168
planters 33, 81, 92, *106,* 120, 121, 133, 135, 137, 151, 153, *157,* 158–9
  see also containers
plants and flowers
  classification of 17
  effect on air pollution 173, 178, 179
  symbolism 29, 37, 45, 65, 166, 182, 197
  see also exotic plants; flower arranging; trade in exotic plants
Plantscape 168
plantscaping
  architects and interior designers relationship 126, 137, 165–6, 168
  on the benefits of nature inside 170, 173

  books 165–7, *166*
  commercialism 173, 176, 177
  history of 11, 22, 130–31, 164, 165, 167–8, 176
  landscape architects relationship 130, 141, 164–5
  plantscaping as a profession 11, 131, 164, 165, 167–8, 178, 179
  training courses 165, 167
Plaza Hotel, New York 64, 148
pleasure gardens 31, 52, *52,* 57, 59, 153, 156
Pohle, Emil 83
Pollini, Gino (Group 7) 78
Portaluppi, Piero 78, *79,* 81
Portman Architects office building, Atlanta *160*
Portman, John 141, 143–4, 146, 148, 149, 156, 183
  The Architect as Developer (with Barnett) 144, *144*
  Entelechy I and II 144–6, 151
  see also Hyatt Regency Hotel
Prairie School style 107
Pritzker, Jay and Don (Hyatt Corporation) 148
private indoor spaces
  eighteenth century 19, 24, 26, 51
  nineteenth century; aristocratic homes 25–7, 29, 33, 37, 40, 63; greenhouses and conservatories *see* conservatories; greenhouses; middle class homes 11, 19, *29,* 29–31, 33, 34, 37, 40, 46
  twentieth century 184; nostalgia influences interior decoration 184–5, *185,* 186, 189; *see also under* Californian modernism; late modernism; modernism
  twenty-first century 168; retro-futurism 186
Propst, Robert 173
proteas 17
'pteridomania' (fern madness) 42
public and semi-public indoor spaces
  nineteenth century 11, 20–22, 51, 65–6; atria 64, *147,* 147–8; convalescent homes 66, *66*; department stores 51, 55, 63, 65; exposed lifts 148; hospitals 66, 175, *176*; hotels 51, *62,* 63–4, *64, 65, 147,* 147–8; mental asylums and sanatoria 66; museums 51, 55; ocean liners 51, 63, 64–5; photographers' studios 66, *67*; private clubs 66; residential schools 66; *see also* botanical gardens; exhibition

buildings; leisure spaces; public parks; winter gardens
twentieth century; hospitals and psychiatric hospitals 163, 173, 175; offices and landscaped offices (Bürolandschaft) 163, 167, *171*, *172*, 173, *174*, 175; prisons 163, 173; restaurants and hospitality spaces *167*, 173, 176–7, *177*; shops and shopping malls 163, *166*, 167, *167*, 173
twentieth-century modernism and late modernism; airships 84, *84*; atria 133, 135–6, 137, 139, *140*, 143, 145, 146–7, 148, 153; interior courtyards see atria *above*; leisure spaces 125; nostalgia influences interior decoration 154; ocean liners 76, 78, *78; see also* hotels; office buildings; restaurants; shopping malls
twenty-first century 189–90; education 168; healthcare 168, 179, 192; hospitality 168, 190, 191; leisure spaces 194, *194*; office buildings 175, 190, *191*, 197; prisons 179; retail spaces 168, 177–8, 179, 191, 196; transport hubs 192, *193*, 194, 196, *196*, 197; wedding venues 9, 192
public parks 22–4, 51, 56

Q
Quickborner 173

R
Rading, Adolf 71, 74, 84
Real Jardín Botánico de Madrid 21
Rees, Yvonne: *Essential English Country Style* 185, *185*
Regent's Park winter garden 54
Reich, Lilly 12, 91, *91*, 94, 99, 100–101, 102
religion and nature inside 23, 37, 49, 65, 103, 186, 197
Re:Mind meditation centre, London (Heath) 168, *169*, 197
Repton, Humphrey 19, 25, 27, 105
Restaurant Associates 127, 128–9, 130
restaurants in twentieth-century late modernism 125, 128–9
see also Four Seasons Restaurant
Riihitie House (Aalto House), Munkkiniemi, Helsinki (Aalto) *80*, 81, 146
Ritz-Carlton Hotel, Montreal 64
Ritz, César 63
Ritz Hotel, London 63

RMS *Berengaria* (Cunard, previously Hamburg America Line) 64
RMS *Olympic* (White Star Line) 64
RMS *Titanic* (White Star Line) 64
Roche, Kevin 131–2, 133, *136*, 136–7
Roderová-Müllerová, Markéta 91
Roehrs family 164
Rohault de Fleury, Charles 22
Roof Arquitectos 189
Roth, Alfred and Emil 76, *77*
Roy F. Wilcox and Company 121–3
Royal Aquarium and Summer and Winter Gardens, Westminster 56, 56–7, 59, 194
cultural and sporting facilities 57, 59
entertainments 59
horticultural shows 59
Royal Botanic Garden, Edinburgh
Temperate House *21*, 22
Tropical Palm House 22
Royal Botanic Gardens, Dahlem, Berlin: Tropical House 22
Royal Botanic Gardens, Kew 16, 21, 22, 41
Palm House 22
Temperate House 22, 183, *183*
Royal Botanic Gardens, Sydney 41
Royal Botanic Society 54
Royal Hotel winter garden, Great Yarmouth *62*
Royal Society 16, 17
rubber plants 12, 35, 69, 83, 101, 112, *113*, 114, *115*, 121, 122, 123, 173, 185

S
Saarinen, Eero 119, 120, 131, 132, 137, 144
Sandcastle House, Malibu (Gesner) 117
Sanssouci palace, Potsdam, Germany: orangery 24, *25*, 73
Savoy Hotel, London 63
Scharoun, Hans *70*, 71, 73, 74, 75, 76
Schindler, Rudolf 105
Kings Road residence, Los Angeles 107
Lovell House, Los Angeles 105–6, 107
Schlitz Hotel, Milwaukee 63–4
Schminke, Fritz 71
Schnelle, Wolfgang and Eberhard (Quickborner) 173
science and the study of plants 15–16, 17, 19, 37, 163
Seagram Building, New York (Mies van der Rohe) 126–7, 129
see also Four Seasons Restaurant
Sefton Park palm house, Liverpool 23, *23*
Selfridges 65
Shapiro, Joel: *20 Elements* 134

Shaughnessy House conservatory, Montreal 26, *27*
Sheats Goldstein House, Los Angeles (Lautner) *116*, 117
shopping malls in twentieth-century late modernism 125, 131, 132–3, 143, 144, 146
see also NorthPark Center
Shrimp house, Fukuyama (UID Architects) 187
Shulman, Julius 109, *110*, 110–11, *113*, 114, *114*, 122
sick building syndrome 173, 178
Siemensstadt residential, Berlin (Scharoun) 75
Sky Garden, 20 Fenchurch Street, London (Gillespies) 194, *194*
Skyview Lounge, Basel-Mulhouse Airport 192
Sloane, Sir Hans 16, *16*
Smith, Harwood K. 133
Smith, Ludwig August: *Interior with Mother and Daughter by a Window 35*, 146
Smith, Whitney R. 111
Smithson, Alison and Peter 117
Somerleyton Hall conservatory, Suffolk 26
Soriano, Raphael 118, *118*
South Africa 17, 21, 23
Southdale, Minneapolis: shopping mall 132, 133
southern California 11, 105, 106–7, 111, 122
see also Californian modernism
Southport winter garden: conservatory 60, *60*
Spence, Benjamin Edward 23
Springer, Marvin 133
Spry, Constance: *How to Do the Flowers* 85
SS *Amerika* (Hamburg America Line) 64
SS *Bismarck* (Hamburg America Line) 64
SS *Imperator* (Hamburg America Line, later RMS *Berengaria*, Cunard) 64
SS *Johann Heinrich Burchard* (Hamburg America Line, later *Reliance*, United American Line) 64–5
SS *Majestic* (White Star Line) 64
SS *Nieuw Amsterdam* (Holland America Line) 65
SS *Normandie* 76, 78, *78*
SS *Reliance* (United American Line, previously Hamburg America Line) 64–5
steel

in Californian modernism 107, 111, 118, 120, 121
in late modernism 135
modernist furniture 92
modernist houses 70, 88, 90, 91
Stone, Edward Durrell 150
Stourhead, Wiltshire
indoor planter by Chippendale 33
Temple of Apollo 18
succulents 17, 94, 117, 123, 196
Sullivan, Louis 71, 154
Sunderland Museum 55
Suppose Design 187
Sutcliffe, G. Lister: *The Principles of Modern House Construction* 38–9
Swiss Banking Corporation offices, Basel 172
Swiss cheese plants 12, 112, *113*, 114, *115*, 117, 118, *118*, 122, 123, 173, 185
Syon House, near London: conservatory 25

T
Tackett, LeGardo 120
Terceira House, Aigburth Drive, Liverpool *36*
Tifferet, Sigal 177–8
Tivoli Gardens, Copenhagen 153, 156
Todd, Dorothy 70
Torquay winter garden 60
trade in exotic plants 11, 12, 16–17, 20, 21
for food and medicines 15, 16, 17, 20
in nineteenth century 16, 19–20, 29–30, 37, 41, 55
for plant collections of the wealthy 17, 19
for scientific research 15–16, 17, 19
transportation of 19–20
trees 9, 109, 114, 118, 123, 127, 128, 187–9, 191, 192
see also fig trees; maples; umbrella trees
Tryg Insurance headquarters, Copenhagen 197
Tugendhat, Fritz and Greta 88, 90, 91, 94, 99, 100, 101–2
tulips 33
Turner, Richard 22, 54

U
UID Architects 187
Ulrich, Roger S. 171, 175
umbrella trees *139*, 173
see also Australian umbrella trees
urbanisation 10, 11, 22

V
Vanderbilt mansion conservatory, 640 Fifth Avenue, New York 26–7
Vauxhall Gardens 52
Victoria Park palm house, Hackney, London 23
Victoria, Queen 54
Vienna International Exhibition (1873) 56
Villa E-1027, Roquebrune-Cap-Martin, France (Gray and Badovici) *72*, 73
Villa Necchi, Campiglio, Milan (Portaluppi) 78, *79*, 81
Villa Noailles, Hyères, France (Mallet-Stevens) 75
Villa, Schminke, Löbau, Germany (Scharoun) *70*, 71, 73, 75, 76
Villa Tugendhat, Schwarzfeldgasse 45, Brno (Mies van der Rohe) 73, 85, 87–95, *88*, 99–103, 149
colours and fabrics 91, 92, 97, 99, 101
designed from inside out 94, 97, 99, 103
exterior 89, 90, 92; terraces 89, 92, 94
furniture 90–91, 94, 95, *95*, *96*, 97, 99, 100, 102
materials used 88, 89, 91, 94, 99, 102
nature inside and out 92–3, 102, 103, 187; cut flowers 90, 95, 99–100, 102; garden 88–9, *89*, 91, 94, *98*; natural setting important 88–9, 90; potted plants 90, 92, 93, *93*, 94, 97, *98*, 102; winter garden *89*, 89–90, *95*, 97, *98*, 99, 101–3
open living space 88, 90, 93, 95, 99, 103
plans and perspective views 94, *98*
rooms and living space; living, dining and study areas 95, *95*, *96*, 97, 100; upstairs rooms 93–5, 100
vestibule 92
technological innovation 88, 90
Vilnai-Yavetz, Iris 177–8
vines 137, 150, 151, 153, 158–9

W
Waldorf Hotel, London 63, *64*
Walker, Rodney A. 118
Waltonian cases 46
Ward, Nathaniel Bagshaw 19–20
Wardian cases 19–20, *20*, 35, *37*, 46
Waring, Samuel J. 31, 40
see also 'Palmyra'
Warrington cases (for ferns and aquaria) *34*, 35
Wave House, Malibu (Gesner) 117

Webbs' *Bulb Catalogue* 43
wedding of Catherine Middleton and Prince William, Duke of Cambridge 9, 192
Weissenhof Estate 71, 74, 84
Went, Frits 139
Wenzgasse 12 residence, Vienna (Frank) 76
Wilk, Christopher 81, 83
Williams, E. Stewart 119, *119*
Williams, Henry T. 29, 30–31, 38, 48
Window Gardening 46–7
Wilson, Edward O.
Biophilia 169–70
The Biophilia Hypothesis (with Kellert) 170
Wilson Street, London: office building 197
window gardening 12, 33, *34*, 35, 37, *37*, 42, *44*, 45, 46, 48, 179
Winter Garden Atrium, World Financial Center, New York (Pelli and Balmori) *191*, 191–2
winter gardens
in nineteenth century 31, 51–2, 54, 55, 56, *58*, 59–60, 66, *66*, 194; other facilities 54, 57, 59, 63; seaside winter gardens 60, 60–61, *61*, 63; social function 52–4, 56–7, 63
in twentieth century; in modernist buildings 73, 76, *77*, *89*, 89–90, *95*, 97, *98*, 99, 101–3; virtual disappearance 69
in twenty-first century *191*, 191–2
Wlach, Oskar 76
Wolverton, Bill 178, 179
women *see* gender issues
Wright, Frank Lloyd 9, 71, 73, 81, 105, 107, 143–4, 149, 191
Fallingwater (Kaufmann residence), Pennsylvania 73, 151, *151*
Frank Lloyd Wright Home and Studio, Oak Park, Chicago 75, 83–4
Guggenheim Museum, New York 148
Hollyhock House, Los Angeles 105, 106, *106*
influenced by Japanese flower arranging 83–4
Martin House, Buffalo 84

Y
yuccas 173

Picture Credits

Alamy Picture Library: 24, 95, 108, 110, 116, 118, 119; / Allan Cash Picture Library: 104; / Amoret Tanner: 34; / Bildarchiv Monheim GmbH: 39; / dbimages: 91; / Dorling Kindersley Ltd.: 4; / INTERFOTO: 103; / Keasbury-Gordon Photograph Archive: 31, 107; / Prisma by Dukas Presseagentur GmbH: 42; / Science History Images: 2; / STOCKFOLIO: 105; / Vintage Images: 22; / ZUMA Press: 80
Alamy Picture Library/Alamy Live News/ZUMA Wire: 112
Alamy Picture Library/Heritage Images/Museum of London: 26
Alvar Aalto Museum, Helsinki: 47
Bridgeman Images/Mondadori Portfolio: 99; / Nationalmuseum, Stockholm: 16
Getty Images: 117, 120; / Bettmann: 77; / Bloomberg: 1; / Condé Nast: 78; / Denver Post: 89, 101; / Fox Photos: 50, 93; / Heritage Images: 41; / Hulton Archive: 19; / LIFE Picture Collection: 90, 92; / New York Daily News: 82, 85; / NurPhoto: 114; / Science and Society Picture Library: 28; / Topical Press Agency: 44; / ullstein bild: 48
Getty Images/AFP/RADEK MICA: 51
Getty Images/Contour/Figarophoto: 71
Getty Images/Corbis/VCG: 94
Getty Images/Hufton+Crow/Universal Images/View Pictures: 109
Getty Images/Photo 12/Universal Images: 27
Getty Images/Universal Images/View Pictures: 113
Historic England Archive/Bedford Lemere & Company: 13, 17, 20
J. Paul Getty Trust, Getty Research Institute, Los Angeles (2004.R.10): 65, 66, 68, 69, 70, 72, 75
MART, Fondo Figini e Pollini: 45
Mary Evans Picture Library: 23, 49
Museum of Modern Art, New York/Scala, Florence: 54, 55, 57, 58, 59, 60, 62, 63, 76
Oliver Heath Design: 102
Queen Mary, University of London Picture Library: 30
Studio 13, Helsinki: 106
Tate/Architectural Review: 43
All other pictures courtesy of the author

Picture Sources

p. 2: Sky Garden at 20 Fenchurch Street, London (detail), designed by Gillespies, 2011. Photograph by Steve Tulley, 2015. Alamy Picture Library

p. 6: Ford Foundation Building, New York, designed by Kevin Roche and John Dinkeloo, 1968. Photograph by Dick Lewis, 1979. Getty Images/New York Daily News

p. 8: Sheats Goldstein House, Los Angeles, designed by John Lautner, 1963. Photograph by Gaelle Le Boulicaut, 2010. Getty Images/Contour/Figarophoto

p. 14: Conservatory, King George VI Memorial Park, Ramsgate, Kent, 19th century. Photograph by the author, 2017

p. 32: Palmyra, Aigburth Vale, Liverpool, 1896. Photograph by H. Bedford Lemere. Historic England Archive/Bedford Lemere & Company

p. 50: Crystal Palace, Sydenham, 1911. Getty Images/Science and Society Picture Library

p. 68: Villa Necchi Campiglio, Milan, designed by Piero Portaluppi, 1935. Photograph by the author, 2011

p. 86: Villa Tugendhat, Brno, designed by Ludwig Mies van der Rohe, 1930. Photographer unknown. Museum of Modern Art, New York/Scala, Florence

p. 104: Eames House, Los Angeles, designed by Charles and Ray Eames, 1949. Photograph by Julius Shulman, 1950. J. Paul Getty Trust, Getty Research Institute, Los Angeles (2004.R.10)

p. 124: Ford Foundation Building, New York, designed by Kevin Roche and John Dinkeloo, 1968. Photograph by the author, 2007

p. 142: The Copenhagen Café, Hyatt Regency Hotel, Atlanta, designed by John Portman, 1967. Photographer unknown, 1968. Getty Images/Fox Photos

p. 162: Landscaped office of the Swiss Banking Corporation in Basel, 1970s. Photographer unknown. Alamy Picture Library/Allan Cash Picture Library

p. 180: G-Tower, Incheon, South Korea, designed by HAEAHN Architecture, 2013. Photograph by Inigo Bujedo Aguirre, 2014. Getty Images/Universal Images/View Pictures

pp. 198–9: Temperate House, Royal Botanic Gardens, Kew, opened 1852. Photograph by Marshall Black, 2018. Alamy Picture Library